QUEEN REARING
and
BEE BREEDING

HARRY H. LAIDLAW JR.

ROBERT E. PAGE JR.

QUEEN REARING
and
BEE BREEDING

Harry H. Laidlaw Jr. and Robert E. Page Jr.
University of California, Davis

Queen Mating Apiary in California

Wicwas Press

First Edition

Published by Wicwas Press, 1620 Miller Road, Kalamazoo, MI
49001 www.wicwas.com

Library of Congress Cataloging-in-Publication Data

Laidlaw, Harry Hyde, 1907-
  Queen rearing and bee breeding / Harry H. Laidlaw, Jr. and
Robert E. Page, Jr.
    p.   cm.
  Includes bibliographical references and index.
  ISBN: 1-878075-08-X
  1. Bee culture—Queen rearing. 2. Honeybee—Breeding. I. Page,
Robert E. II. Title
SF531.55.L35 1996
638'.145—dc20   96-27881
                                                            CIP

ISBN: 1-878075-08-X

It has long been known that *the* unique bee of a honey bee colony, successively called General, King, or Queen as biologists and beekeepers learned more about this caste, is indeed a bee of supreme importance. The welfare of the colony, and in fact the survival of the colony, is intimately associated with the queen because of her role as mother of the colony and her influence on worker bee activities through her pheromones. It is evident that the attainment of greatest pleasure and satisfaction of beekeeping by beekeepers depends on the contributions the queen makes to her colony. These contributions are influenced by the genetic composition of the queen and by her physical development and vigor.

Beekeepers and scientists have devised highly successful and economical methods to rear excellent queens. Need for these queens, whether reared by beekeepers for their own use or whether purchased from professional queen breeders, will continue and expand as more frequent requeening becomes necessary to cope with mite diseases and with hazards of Africanized bees. New or revised queen rearing methods are continually appearing and it is desirable to include in this book those that can be most useful to beekeepers.

The identification of DNA *(deoxyribonucleic acid)* as the universal chemical and physical substance of heredity of organisms, and the rapid advancement in understanding and manipulation of elements of the genetic system as revealed by molecular genetics, demonstrate the necessity of at least a layman's grasp of hereditary principles by beekeepers, and constitute a prediction of increasing employment of molecular techniques in studies of honey bee breeding, behavior, and diseases, and in practical applications. To help beekeepers appreciate the importance of such research and how it can benefit them, a second focus of this book is an introduction to honey bee genetics and bee breeding.

Conversely, bee geneticists need reasonable skill in queen rearing and beekeeping to organize and perform research that will result in practical benefit to beekeepers and thus to users of honey bee services and products. They need this skill also when their research is purely investigative.

In the preparation of this book, the authors drew on their own observation and experience, and on papers and books published by them and by others. They relied extensively on *Queen Rearing* by Laidlaw and Eckert, published by University of California Press, and on *Contemporary Queen Rearing*, and on *Instrumental Insemination of Queen Bees*, both by Laidlaw, and published by Dadant and Sons, Inc. Sections of Chapters VII and VIII were extracted and modified from R.E. Page and H.H. Laidlaw (1991) Honey Bee Genetics and Breeding, In *The Hive and The Honey Bee, Joe Graham (ed.)*. Dadant and Sons Inc. Chapter IX was extracted and modified from R.E. Page and E. Gunmán-Novoa (in press) titled: The genetic basis of disease resistance, In *Honey Bee Pests, Predators, and Diseases*. R.A. Morse (ed.), A.I. Root Co. We thank Dadant and Sons, Inc., and A.I. Root, Co. for their permission.

We express our appreciation to Kim Fondrk for conducting the experiment that determined the distribution of foraging bees in the hive after flight had ceased, and for preparing the illustration to show this. Gus Rouse of the Kona Queen Company kindly sent us a current battery box, with directions for its use, to photograph for the book.

We thank Ruth Laidlaw for her patient help in preparing the many typescript versions of this book.

"Written for beekeepers who know little about genetics; and geneticists who know little about beekeeping."

H.H.L.and R.E.P.

Contents

A BRIEF HISTORY OF QUEEN REARING

A review of available beekeeping literature reveals the interest that is shown in the queen bee and in her relation to the colony (see Laidlaw 1979). From the discovery that this complex insect society had a "queen", observers began attempts to find the source of her influence on her vast number of "subjects". Only within the past forty years have we learned the source of that power. For many centuries, fragments of information were assembled on her life history, how she is reared naturally, and how queens of high quality are produced under the beekeepers' control. Essential tools of beekeeping were discovered that made it possible to fully exploit the economic value of honey bees. Without the movable-frame hive, comb foundation, the honey extractor, or the bellows smoker, beekeeping would indeed be at the same stage of advancement it was in the first half of the nineteenth century. A brief summary of some of the discoveries of the early years will illustrate the basis for the development of the advanced systems of queen rearing used at the present time.

Early Knowledge of the Queen

Our knowledge of the life history and behavior of bees began some hundreds of years before the beginning of the Christian era. Aristotle (Haeckel 1896), a Greek philosopher who lived in 384-322 B.C. is given credit for being the first to write a scientific account of the natural history of the honey bee. From his writings we learn that the beekeepers of that day were undoubtedly aware of the presence of a queen bee in the colony. Accustomed to kings and queens or to dictators, people reasoned that a society of insects as highly developed along socialistic lines as the honey bees must also be ruled by some similar directive influence. Columella who wrote during the reigns of Tiberius and Caligula around the middle of the first century of the Christian Era, in a chapter on bees, contributed the thought that the "King of the Bees ... must be stript of his wings when he often makes eruptions with his swarm, and endeavors to run away: for having pulled off his wings, we shall retain the vagabond General as it were, with a chain at his foot: who being deprived of all means of making his escape, will not care to go without the bounds of his kingdom: for which reason he does not indeed allow the people of his dominion to ramble up and down, and wander at a greater distance from him." These early scholars did not know that she was the mother of all the bees in the colony; it was not until centuries later that this simple fact was discovered. Charles Butler (1634), an English bee-keeper, appears to have been the first to bring to public attention, in 1609, that the colony was indeed a *Feminine Monarchy*. He thought it was ruled by a "Queen" that perpetuated the life of the colony by producing daughter queens.

Near the close of the seventeenth century, Swammerdam (1732) studied the anatomy of the honey bee with the aid of a microscope. He made many remarkable discoveries. Among other things, he established conclusively the sex of the queen and the drone. Around 1712, Maraldi (Bevan 1838) devoted considerable time to a study of bee behavior; he is credited with being the first to use glass-sided hives. From 1732 to 1744, Réaumur (Bevan 1838) used this device in efforts to find how the queen mated, but he erroneously concluded that she mated with the drones while in the hive. Bonnet, of Geneva, rediscovered the phenomenon of par-thenogenesis in 1740, when he showed that plant lice or aphids could reproduce without fertilization; and others were led to search for the same method of reproduction in other insects.

Anton Jansha (1771), in studying swarming of honey bees, introduced swarm cells and virgins of afterswarms to small, newly made-up colonies (Alphonsus 1931). Such colonies are now known as "increase".

Also in 1771, Schirach (1787), contemporary with Jansha, was able to prove conclusively that queen bees can be reared from larvae in worker cells. He seems to have been the first beekeeper to use small hives for the rearing of queens, and to him might also go the recognition of being the first to use baby nuclei. Schirach demonstrated that queens can be reared from worker larvae by shutting up a "handful of bees in a box with a small piece of comb containing eggs and maggots in worker cells." He kept the bees confined for two days, at the end of which they were found to have modified a number of the worker cells into queen cells.

While these discoveries were being made, Francois Huber (1814), a blind Swiss naturalist, inspired by the memoirs of Reaumur, undertook to study bee behavior with the efficient aid of his wife and a very intelligent and able secretary, Francis Burnens. Huber wrote accounts of his experiments in letters to Bonnet in 1791, who gave him every encouragement. Huber was able to prove that queens mated only on wing, and when so mated could lay either worker or drone eggs at will. He showed that only drones were produced when queens were confined to their hives, without mating, until they began to lay. He confined the queens by restricting the entrance with a glass tube that permitted workers to pass through but not drones or queens. This finding confirmed the observations of Aristotle and others. Huber also repeated the experiments of Schirach and improved on them to the extent of demonstrating that queens can be produced from very young larvae instead of from larvae that were "three or four days old." In these experiments, Huber and his helpers placed combs containing eggs and very young larvae in queenless colonies and had them start queen cells, as had been described by Schirach. But, to prove that queens could be reared from young larvae, he had the larvae removed from the queen cells and replaced with others that were known to be only 48 hours old. Two queens emerged from five of the queen cells thus produced. In this simple experiment, Huber not only demonstrated that the bees could produce queens from young worker larvae but in doing so used the method of transferring larvae from worker cells to queen cells which is now used almost exclusively to rear queens commercially, the only difference being that he transferred larvae to natural queen cells rather than to artificial queen cell cups. He also invented the leaf hive, one of the first hives in which the combs were movable, in order to facilitate his observations on the life history and habits of bees. As a means of increasing the number of his colonies, Huber divided his leaf hives and permitted the queenless portions to rear their own queens. He apparently did not realize, though, the full significance of the possible use that could be made of the method of rearing queens he had demonstrated in his experiments.

The discoveries of Schirach and Huber contributed the way for the production of queens on a limited scale, even when most of the hives were of the immovable-frame type. Their methods persisted, with few variations or improvements, for the better part of a century.

Parthenogenesis Discovered in the Honey Bee

The next important advance in knowledge of queens was made by Dzierzon who in 1845 published in the *Eichstadt Beinen-Zeitung* his theory that drone bees are produced without the eggs being fertilized. This observation appeared one hundred years after Bonnet discovered parthenogenesis in aphids. Dzierzon also published a book, *Theory & Practice of Bee Culture*, in 1848. He was more fortunate than Huber and some of his other predecessors in that he had two races of bees of different colors to experiment with in the final proof of his theories. He used not only the black bees of Silesia but also the Italian or yellow bee, which he introduced into Germany. It was the introduction of the latter race that enabled him to convince beekeepers that drone bees came from the unfertilized eggs of queens. When his unmated Italian queens as well as Italian queens which had mated with black drones could produce only drones which were true to the Italian race, he reasoned that all of the eggs produced in the ovaries of queens were unfertilized and that the sex of the individual bee depended on whether the egg was fertilized or not as it passed through the vagina at the time it was laid. The production of drones by egg-laying workers was additional evidence for the correctness of his theory.

Further proofs to substantiate his theories were produced by Professors von Siebold and Léuckart (Berlepsch 1861; von Siebold 1857; Léuckart 1861) who demonstrated that sperms were present in eggs that were laid in worker cells by a fecundated queen but were absent in eggs laid by the same queen in drone cells. Léuckart gave a Miss Jurine credit for having first proved by dissection in 1813 that worker bees are females. Baron von Berlepsch (Berlepsch 1861) did much to confirm Dzierzon's theories and collaborated with von Siebold and Léuckart in their studies. These theories did not meet with ready acceptance by beekeepers but are today undisputed.

Dzierzon devised a bar hive in which combs built on bars could be withdrawn horizontally from the back of the hive. His observations and methods of colony management and of queen rearing had a profound effect on beekeeping throughout Europe.

Samuel Wagner (Phillips 1928) translated from the German into English Dzierzon's "Theory or the Fundamental Principles of Dzierzon's System of Bee-Culture" and had the translations printed in the *American Bee Journal* for wider distribution.

Early Progress in America

While these developments were taking place in Europe, progress was being made in the United States along almost parallel lines in many phases of the industry (Pellett 1918). In reviewing the progressive developments in beekeeping, it is necessary to remember that the use of the movable-frame hive, comb foundation, the honey extractor, or even the bellows smoker came relatively late. Those who have never seen colonies handled without the aid of these essential tools may have difficulty in realizing the handicaps under which the early beekeepers operated or the changes which the discoveries have brought. Each development came only after many years of trial-and-error efforts by many individuals in widely separated localities, and usually without the benefit of knowing what others had accomplished. Dzierzon and Langstroth, for example, were working on the development of a movable-comb hive without being aware of each other's work, and the results of their discoveries advanced the interests of beekeeping in a few short years more than had been achieved in centuries before their time.

Queen rearing in the United States began by beekeepers using swarm cells for requeening, and for colony replacement and increase. Moses Quinby, who is considered by many to be the father of commercial beekeeping in the United States, published the first edition of *Mysteries of Beekeeping Explained* in 1853, apparently without knowing of the work of Dzierzon or of Langstroth. Quinby advocated the use of swarm cells, cut from colonies making preparations to swarm, as a means of requeening those which had lost their queens or for providing queens for colonies made by division or by artificial swarming. He described how queen cells could be built by queenless colonies if they were given small pieces of comb containing eggs or young larvae from queenright colonies, but wrote that he had better success with cells built under the swarming impulse or in stronger colonies. He made no contribution to the advancement of queen rearing.

In the same year, Rev. L. L. Langstroth published a description of his movable-comb hive and of his system of colony management under the title *Langstroth on the Hive and the Honey-Bee.* Without doubt, this was the most constructive publication on beekeeping written in the English language until that time. Langstroth not only provided a description of the most successful movable-comb hive that has ever been invented and an outline of its advantages, but also described in detail how colony increase could be made without the benefit of natural swarms. His early method of rearing queens was to cut out sealed queen cells from colonies made queenless for that purpose, or cells produced under the swarming impulse. He used nuclei in divided hive bodies, each nucleus consisting of one frame of brood and adhering bees. If a sufficient number of bees did not adhere to the comb, he shook additional bees into the compartment until they clustered over the brood, or he placed the nucleus in the location of the parent colony until a sufficient number of bees had entered. The nuclei were then given water and were kept closed for one to three days, after which each nucleus was given a sealed queen cell about ready for emergence. Langstroth called this method his *nucleus* system of making increase or of producing queens.

The importation of Italian bees into the United States greatly stimulated interest in beekeeping and queen rearing. Wagner, Langstroth, and Colvin received a shipment of seven queens from Dzierzon in 1859, but these were said to have died the following winter after only a few queens had been reared from them. Parsons (Quinby 1873), an agent of the U.S. Division of Agriculture, is also reported to have received a shipment of Italian bees from Italy in 1859. In 1860 he made some of his stock available to Cary, to Langstroth, and to Quinby, the last of whom began to produce Italian queens in 1861. Additional shipments of Italian bees were received by various beekeepers after that date and Italian queens were soon offered for sale by a number of queen breeders.

Shipping Queens by Mail

Other developments were gradually raising queen rearing to commercial status. The first shipment of a live queen by express was made in 1863 from C. J. Robinson (Quinby 1873) in New York to Langstroth in Ohio. Who made the first shipment of queens by mail is not entirely clear, but the event certainly occurred within the same decade. The early shipments were made in small screened boxes; pieces of comb honey were enclosed for use as food by the bees and queen. Leakage of honey and injury of a postal clerk by a sting through a cage screen prompted a ruling by the postal authorities in 1872 to exclude bees from the mails (Quinby 1873). Shipment of queens by mail did not stop entirely, however, for the ruling was not enforced at all post offices. In 1873, A.I. Root and others shipped by mail pieces of combs containing eggs from Italian queens. Quinby successfully mailed a queen to the *American Bee Journal* in a cage provisioned with lump sugar and a wet sponge. Henry Alley wrote in 1873 that he had been shipping queens successfully for six years; he supplied food by soaking a sponge in dilute honey and squeezed the sponge sufficiently to prevent any drip.

By 1878, "Dollar Italian Queens" were being advertised in *Gleanings in Bee Culture* by thirty-two beekeepers. Some queen breeders were using a water-jacketed box heated by a lamp to incubate their queen cells. A nursery cage for the cells had been invented and was used by some breeders to secure virgin queens for introduction into their nuclei.

I.R. Good in 1881 advocated the use of a queen-cage candy made by mixing sugar with cold honey until firm enough to prevent any drip. Frank Benton perfected a mailing cage in 1883. A.I. Root in 1884 quoted a letter from Benton indicating that the latter had shipped queens successfully in cages provisioned with a candy made from "pounded" sugar and honey. Root also stated that he had made queen-cage candy of powdered sugar and honey; he called it "Good" candy because I.R. Good had first called attention to the value of using a combination of sugar and honey for shipping bees and queens. Langstroth, as early as 1859, had given in *The Hive and the Honey Bee* a formula made by a Rev. Scholz for feeding bees in winter. It recommended mixing one pint of warmed honey with four pounds of pounded lump sugar until a stiff doughy mass was formed. Apparently this formula escaped the attention of American beekeepers interested in shipping queens until after Good had recommended a combination of sugar and honey for that purpose.

The American version of this queen-cage candy is still known as "Good" candy. At present, it is made by mixing powdered sugar with an inverted sugar syrup or a commercial inverted sugar syrup called Nulomoline™. It may also be made with powdered sugar or Drivert™ and high fructose corn syrup. The combination of the Benton cage and the use of a soft candy that did not stick up the bees or leak from the cage provided a fairly satisfactory solution of the problem of shipping queens by mail.

Improved Methods of Cell Building

Interest in imported bee races and in rearing queens for home use and for sale continued to grow in the 1880s accompanied by new and modified methods of rearing. The first to use the term *"grafting"* to refer to the substitution of worker larvae for those found in naturally built queen cells was E.C. Larch in 1876. In this, he duplicated the method first used by Huber, which is essentially double grafting. J.L. Davis (Davis 1874) has also been credited with having used the method in 1874, when it was referred to as the "Davis Transposition Process."

W.L. Boyd in 1878 suggested the feasibility of cutting out and saving naturally built queen cell cups into which newly hatched larvae could be transferred and then placed in queenless colonies. A.I. Root (Boyd 1878) improved on this idea by suggesting the possibility of using wooden cups, to hold artificially built queen cell cups into which larvae could be grafted, or into which the "whole bottom of the cell" could be transferred with the larva. Root noted that the idea of making the cell cups artificially was suggested by "somebody" previously.

O.H. Townsend (1880) advocated the production of queen cells by cutting worker comb containing young larvae or eggs into narrow, one-cell-wide strips and fastening them to the surfaces of combs with the cells pointing downward. He placed the combs, with the strips fastened near the top bars, in queenless colonies and removed larvae from some of the cells after the bees had partially drawn out the cells. He said that he seldom left more than twenty cells at a time in each cell-building colony. J.M. Brooks (1880) improved on this method in the same year. He trimmed the strips of cells containing the eggs to one-quarter inch of the midrib, then fastened the strips to wooden bars which he called "cell bars" and placed them in frames for queenless colonies. The queen cells were then built from the cells containing the eggs or larvae.

Henry Alley advanced beekeeper control of queen rearing yet another step when, in 1883, he improved the method of getting queen cells started. He introduced a "swarming box", and some other queen rearing manipulations that are worthy of notice. His breeding queens were confined to small hives that had five frames, each approximately five inches square. The center frame contained worker comb in which brood had been reared on one or two occasions. The other four frames were left for brood and honey. A good queen would completely fill the center comb within 24 hours after it was placed in the hive. This comb was then removed and placed in a queenless colony for incubation and first larval feeding. When the eggs started to hatch, the comb was cut into strips having a center row of cells intact. The eggs or larvae in alternate cells in this row were destroyed and the cells cut down with a thin, warm knife, to one-quarter inch of the midrib. These strips were then fastened with melted wax and rosin to brood combs which had the lower half cut away on a slightly convex curve, a procedure that spread the cells and prevented their being built together.

To establish the cell-starter swarm boxes, the bees from a number of colonies were shaken into "swarming boxes" and held in a cool, dark place for ten to twelve hours. The brood from the colonies from which the bees were taken was divided between weaker colonies. At the end of the day the bees from the swarming boxes were transferred to broodless cell-building hives located on their original stands, and each was given a prepared comb with a strip of cells. The cell-building colonies were fed sugar syrup. When the cells were completed, they were either distributed to small four-frame nuclei or stored in nursery cages in an incubating colony. The cell builders were given the combs of brood from the next lot of colonies used to fill the swarming boxes. Alley stocked each small "fertilizing or miniature hive" with a pint of bees and kept the bees confined for two to three days in a cool, dark room after which they were placed on location. The miniature hives were opened at nightfall and given a ripe cell or a virgin queen. Sugar syrup was fed to each nucleus from an internal feeder that was supplied with syrup through a hole in the cover.

Alley also made an improved nursery cage, and he wrote on the care and introduction of queens as well as on many details pertaining to queen rearing.

The production of queen cells from strips or plugs of worker comb containing newly hatched larvae captures the interest of beekeepers even today. The queens produced are good if the larvae are prepared when a day old or younger and are fed to sealing in a strong, well-maintained cell builder colony. We do not know of evidence that the queens are superior, though Smith (1949) claimed they are. Use of alternatives to grafting is a matter of personal preference.

The Doolittle System of Queen Rearing

G.M. Doolittle became interested in rearing queens in about 1870. During the next eighteen years he gradually developed an improved system while testing all of the then-known methods. Doolittle was a keen observer, and, while he did not originate many new principles of queen rearing, he showed great ingenuity in combining the advantages of numerous methods advocated by others into a workable system. He wrote intensively for the bee journals, describing his methods of queen rearing and beekeeping.

Doolittle published *Scientific Queen-Rearing* in 1888 in which he described his experiments and the method he finally devised from the recommendations of others as well as from his own observations. This book, recently reprinted in *BeeScience*, may still be read with profit by anyone interested in rearing queens.

Doolittle used the method of making artificial queen cell cups and the technique of transferring young worker larvae into them. He attached twelve of the cell cups to each bar of "frame-stuff" (a portion of a bottom bar) and fastened these bars into a frame, the upper portion of which contained comb. A small amount of royal jelly was placed into these cell cups. Worker larvae less than 36 hours old were then transferred into them. He experimented widely with different types of cell-building colonies and finally had the cells built in the second story of a strong colony in which the queen was confined to the lower chamber by an excluder. This is now the typical queenright cell builder and cell finisher.

When the cells were ten days old, they were placed in nuclei or in nursery cages. He experimented for a while with small nuclei whose frames were about five by six inches, but finally decided on two or three frame nuclei made up with regular-sized frames. By these methods he demonstrated that queens could be reared efficiently in quantity. This work established Doolittle's position as the founder of commercial queen rearing.

Refinements

Subsequently, some refinements have been added to the techniques of queen rearing. Frank C. Pellett, then field editor for the *American Bee Journal*, published *Practical Queen Rearing* in 1918 which summarized much of the best information available up to that time. Five years later, Jay Smith in *Queen Rearing Simplified*, gave his experiences in the use of the Doolittle system of producing queens. R.E. Snodgrass's monumental textbook, *The Anatomy and Physiology of the Honeybee*, appeared in 1925. While it did not contribute to queen rearing it was the basic source of information in the development of instrumental insemination of queen bees.

instrumental insemination

Dr. Lloyd R. Watson (1927) first demonstrated his method of artificial insemination of the queen in 1926. The goal of artificial insemination research is to control the queen's mating completely in order to study inheritance in honey bees and to successfully breed them. Watson's skill in designing and making micromanipulators, and in their use, led him to refer to artificial insemination of queens as *instrumental insemination*—a designation that has persisted. Since then, others, notably Nolan (1932), Laidlaw (1934, 1939, 1944, 1949, 1977), Mackensen (1947, 1951), and Mackensen and Roberts (1948) established instrumental insemination as a reliable and efficient control of queen mating. Harbo (1985) contributed a large capacity syringe that has gained acceptance.

Mackensen and Tucker (1970), Laidlaw and Goss (1990), Kühnert and Laidlaw (1994), F. Ruttner, and P. Schley have contributed further variations of instrument or method.

The package bee industry has grown steadily. It now provides hundreds of thousands of colonies each year for pollination and for honey production. This development has encouraged refinements in the movement of queens and bees by mail, express, truck shipments, and by air, and made possible transportation of large movements of colonies over long distances for crop pollination.

THE QUEEN

The honey bee queen is a reproductive specialization of the female sex of honey bees. Her ovaries are very large, occupying most of the abdominal cavity. The reproductive tract, including the spermatheca, is highly developed. Her secretions identify her as a queen, and are recognized by workers, drones, and other queens. Certain physical features of worker bees are suppressed, such as pollen baskets and wax glands, as are worker instincts of brood care, comb building, foraging, and others. Navigational ability has clearly been retained. Her animosity toward other queens is intense so that more than one queen in a colony is rare.

Relation of the Queen to the Colony

The role of the queen in the colony is a very basic one indeed. She is (1) the mother of all members of the colony that develop during her reproductive life. Her value to the colony—and the value of the colony to man—depends on her ability to maintain a population of workers and drones adequate to perform all the functions necessary to its existence as a community of insects. The queen is (2) the breeding depository of all of the inherited characteristics of the colony acquired through her progenitors and through the acquisition of the sperms from the males at the time of mating. Thus, the queen is responsible for the color of the bees, their industry, degree of gentleness, resistance to disease, swarming tendencies, longevity, comb-building propensities, and for many other colony and individual attributes. By merely changing the queen in a colony, one can change many of the above characteristics within a few weeks.

A mated queen is (3) capable of laying unfertilized eggs (Fig. 1), that produce males and fertilized eggs that produce females, thereby controling the sex of her offspring. Workers and queens are females; drones are males. Eggs destined to produce females are laid in worker or queen cells; those that are to develop into drones are laid in drone cells. Unfertilized eggs usually produce

Figure 1. Eggs of the queen bee (magnified). From *Contemporary Queen Rearing*, Dadant and Sons, Inc.

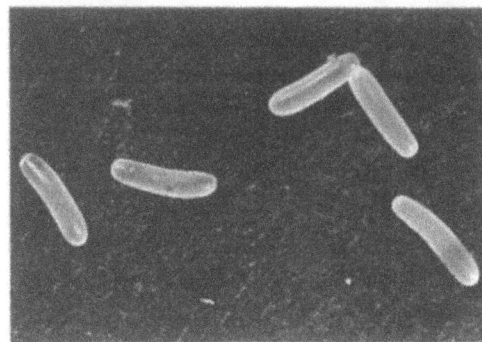

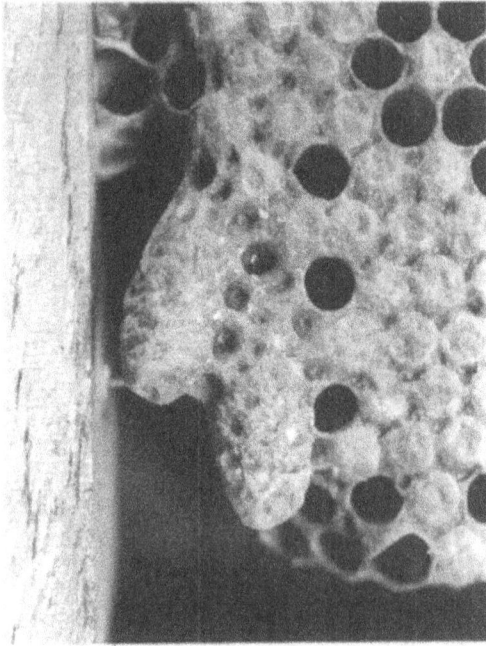

Figure 2. Sealed natural queen cells.

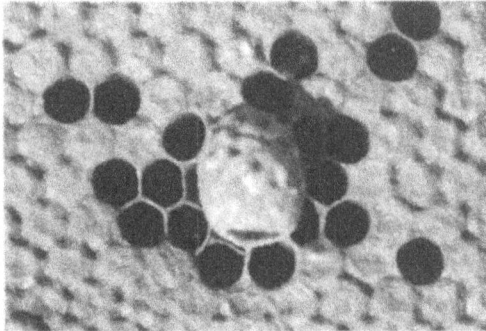

Figure 3. Preconstructed natural cell cup. From *Contemporary Queen Rearing*, Dadant and Sons, Inc.

only drones, but all female larvae can develop into queens or workers, depending on the food and care given them during their early larval stages. Queens originate (4) all honey bee genomes: those of eggs and those of sperm.

Though the queen is the mother of a colony, she lacks "maternal instinct" and is not the sole controlling influence of the work of the colony. She may lay all of the eggs, but the number of eggs she lays depends on the amount and kind of food she receives from the nurse bees. The workers provide a laying queen with her food requirement when she needs it, feeding her on a mouth-to-mouth basis; in this way controlling the egg-laying activities of the queen and correlating the activities of the colony with the season, the weather, and the floral and stored food conditions.

The queen's presence is signaled by her pheromones that she involuntarily produces. The pheromones motivate (5) the workers to maintain the coherance, or integrity, of the honey bee population as a familistic colony, and influence the "morale" of a colony; the behavior of a queenright colony is considerably different from one that is queenless.

The queen provides no maternal care for the eggs she lays and none for the resulting larvae; worker bees take over maternal duties for all the brood after the eggs are laid. The queen is, therefore, an animated egg-laying machine intimately regulated by environmental conditions within and without the hive. But she is the most important single bee in the hive; without her, or sufficient material from which to rear other queens, the colony is doomed to extinction.

Development of the Queen

In the natural course of colony activities, queens are reared only under three major circumstances: queenlessness, swarming, and supersedure. All larvae that hatch from the eggs laid by the queen in worker cells are fed the same kind of food during the first two days of their existence (The exception is diploid drones in worker cells. See chapter VII for discussion of diploid male production). Thereafter, they are fed in a different manner and with a different quantity and quality of food depending on what they are destined to become. The fertilized egg is genetically female, but is neither queen caste nor worker caste; the larva that hatches from a fertilized egg can develop into either caste, that is, into workers or queens. Drone larvae, of course, cannot develop into anything but drones.

Queens are reared in queen cells which extend downward from the face or edges of the comb (Fig. 2). The queen cells may be started as cell bases in advance of the deposition of the egg or may be built around a cell containing an egg or larva. Those built in advance usually begin as cell cups that look something like a small acorn cap (Fig. 3). Those built around worker cells are simply enlargements of the worker cells (Fig. 4). Park (1949) refers to these two types of queen cells as *preconstructed* and *postconstructed* queen cells.

In an emergency, as when a laying queen is killed during the manipulation of the hive, the bees will construct a number of queen cells around larvae in worker cells (Fig. 4). These are built over a period of three to five days, and will contain larvae in different stages of development. When a queen cell is started in worker comb the amount of royal jelly is increased to such an extent that the larva is floated out of its cell into the enlarged cell. Frequently, such cells are started when a colony is being requeened and while the new queen is still confined to her cage, or when queens are purposely confined to control brood rearing.

A colony preparing to swarm builds queen cell cups on the bottom or sides of combs (Fig. 5). The queen lays in some of the cell cups, and additional queen cells may be constructed around cells in worker comb that contain eggs or larvae. One often finds several queen cells in varying stages of development, from newly sealed cells to those recently started. The number of cells started

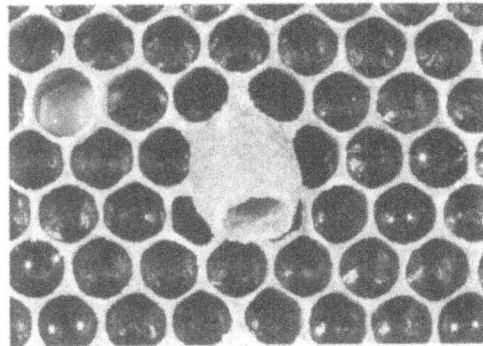

Figure 4. Postconstructed queen cells being built from larvae in worker comb.

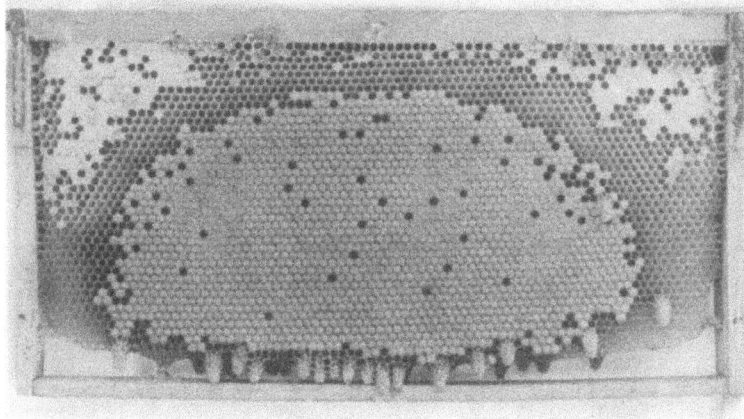

Figure 5. Swarm cells built along bottom of brood comb.

is variable; and some strains of bees build more cells than others. Swarm cells may also be found on the surface of the comb located in comb imperfections.

Aged or injured queens are superseded. The bees build queen cells to replace the disabled queen, generally not more than two or three along the lower and end edges of a comb, and some-

Day	Workers Stages	Workers Moults	Queens Stages	Queens Moults	Drones Stages	Drones Moults
1						
2	Egg		Egg		Egg	
3						
4	1st larval	(hatching)	1st larval	(hatching)	1st larval	(hatching)
5	2nd larval	1st moult	2nd larval	1st moult	2nd larval	1st moult
6	3rd larval	2nd moult	3rd larval	2nd moult	3rd larval	2nd moult
7	4th larval	3rd moult	4th larval	3rd moult	4th larval	3rd moult
8		4th moult		4th moult (sealing)		4th moult
9	Gorging	(sealing)	Gorging		Gorging	
10	Pre-pupa		Pre-pupa	5th moult		(sealing)
11						
12		5th moult			Pre-pupa	
13			Pupa			
14						
15	Pupa					5th moult
16			Imago	6th moult (emerging)		
17						
18					Pupa	
19						
20		6th moult				
21	Imago	(emerging)				
22						
23					Imago	6th moult
24						(emerging)

Table 1. Length of developmental stages of the honey bee. Modified from Bertholf, L.M. 1925. The moults of the honeybee. *Journal of Economic Entomology* 18(2): 380-384.

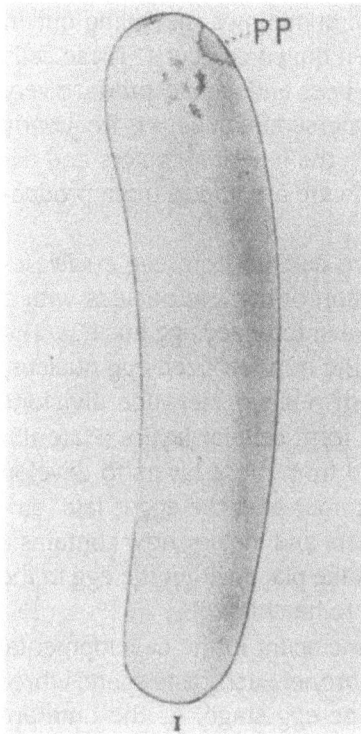

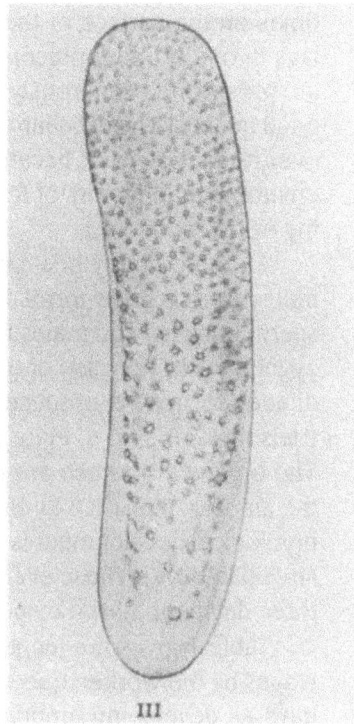

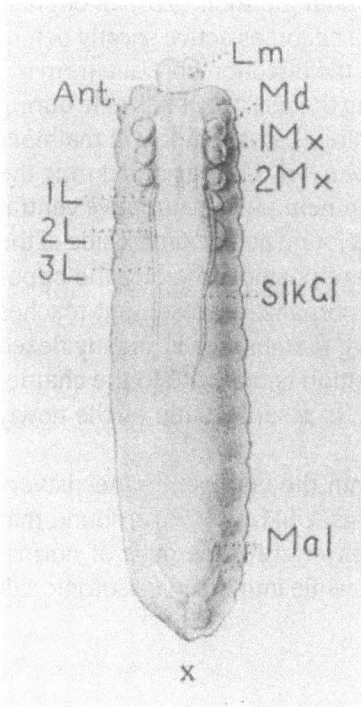

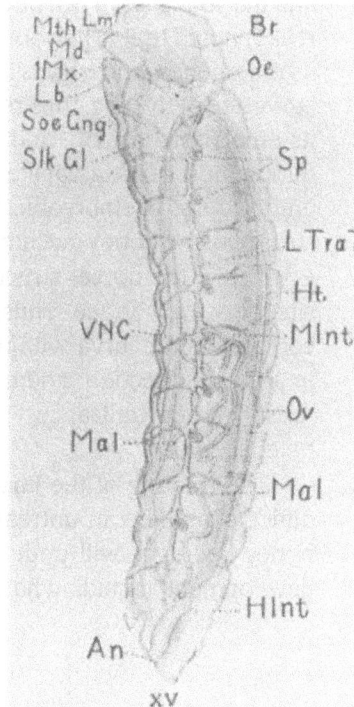

Figure 6. I. Beginning of nuclear cleavage in newly laid egg. III. Cleavage nuclei moving in egg protoplasm to form germ layers. X. Initiation and forming of larval organs. XV. Larvae at end of embryonic development and just prior to hatching. From Nelson, J.A. 1915. *Embryology of the Honey Bee*, Princeton University Press.

Label key

An	anus
Ant	antenna
Br	brain
Ht	heart
HInt	hind intestine
1L	fore legs
2L	middle legs
3L	hind legs
Lb	labium
Lm	labrum
LTraT	longitudinal tracheal trunk
Mal	malpighian tubules
Md	mandibles
MInt	mid intestine
1Mx	first maxillae
2Mx	second maxillae
Mth	mouth
SlkGl	silk gland
SoeGng	suboesophageal ganglion
Sp	spiracle
Oe	oesophagus
Ov	ovaries
PP	polar protoplasm
VNC	ventral nerve cord

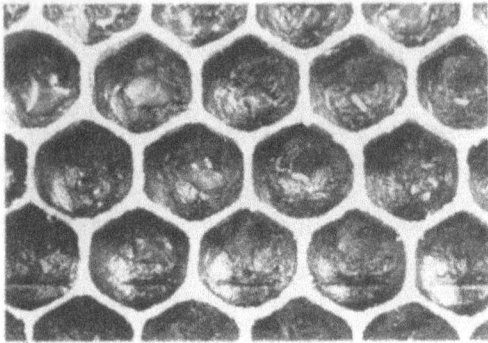

Figure 7. Hatching eggs and newly hatched larvae.

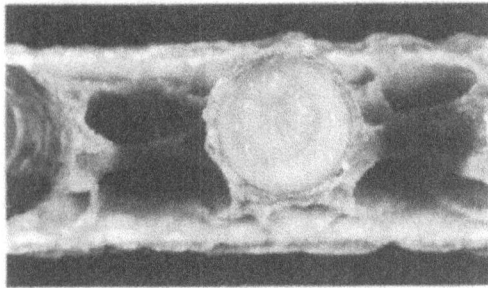

Figure 8. Well fed 4-day old larva in grafted queen cell. Third day from grafting.

times on the surface of the comb. Sometimes the failing queen lays in one or more preconstructed queen cell cups. These cells are provisioned generously by the bees and usually produce very good queens. The excellence of supersedure queens is frequently over rated, however, because both the inherited factors and the environmental factors of food and care are important in producing excellent queens.

All three honey bee castes are derived from egg *nuclei,* females from *zygotes* formed by union of the egg nucleus with a sperm nucleus, and males from the unfertilized egg nucleus. The zygote, (or in the case of drones the non-fertilized egg nucleus) divides to initiate a succession of nuclear cleavage divisions (Nelson 1915, Fig. 6, Plate I) that form cellular layers (Plate III). The organs of the bee are derived from these layers to develop the *embryo* (Plate X). By the 76th hour after the egg is laid, embryological development is complete and the egg now contains a *larva* (Plate XV). These events all take place within the egg in the three day period from oviposition to hatching.

Table 1 gives the length of time spent in the developmental stages by the worker, queen, and drone. Each caste spends three days as developing embryos in the egg stage. As the time approaches for the larva to hatch, the chorion becomes transparent and the young larva can be seen within (Bertholf 1925). Snodgrass (1925) noted that "The young larva becomes active shortly before its emergence, and curves itself in the direction opposite from the curvature of the egg, thus reversing the position it has held during its embryonic growth, and assuming that which it will maintain during most of its larval life." In the newly laid egg and during the entire course of embryonic development, the presumptive ventral surface of the honey bee embryo is found at the convex side of the egg, while the dorsal structures are positioned along the opposite, concave surface. Thus, as embryonic development reaches completion the larva within its egg membranes is slightly flexed in a dorsal direction, and this position is opposite to the characteristic strong ventral curving that is assumed later by the newly hatched larva.

The rotating of the larva within the egg membranes is very difficult to detect in untreated eggs. DuPraw (1961a) found that honey bee eggs will undergo their entire three days of normal development, including hatching, while immersed in isotonic salt

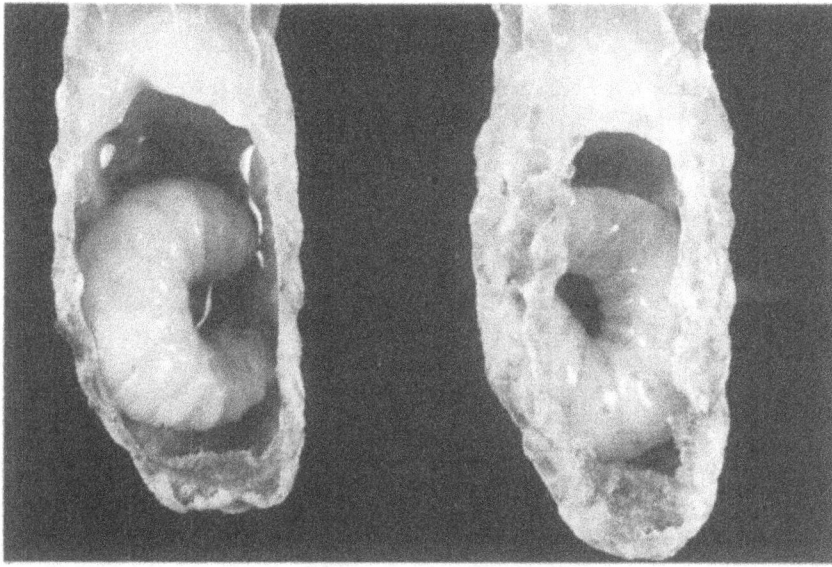

Figure 9. Five-day old larvae gorging on food in sealed queen cells. Four days from grafting.

solution or other liquid media. Under these conditions, the events occurring within the egg are easily visible under a low-power microscope.

DuPraw (1961b) found also that the young larva produces a secretion that dissolves the chorion starting at the top and continuing downward toward the attached base. The larva bends until the head touches the vertical bottom of the cell (comb cell-sides are horizontal), by which time the chorion is almost completely dissolved. Shortly before this hatching process, the nurse bees deposit a supply of thin jelly on the bottom of the cell (Fig. 7), and after hatching, the larva adheres by its side to its food supply. The nurse bees then mass-feed jelly to the larvae for two days after which worker and drone larvae are fed at intervals while queen larvae receive a surplus of food (royal jelly) in every well-cared-for cell (Fig. 8). At the end of four and one-half to five and one-half days from larval hatching the bees make the final cell provisioning and seal the queen cell. The queen larva continues to gorge (Fig. 9) on the abundant surplus food and to gain in weight for approximately 12 hours (Oertel 1930). The time of sealing the queen cell may vary a few hours with different larvae of approximately the same age.

During larval development, the queen larva sheds its skin, or

Figure 10. Six-day old larva spinning cocoon. Five days from grafting. From *Contemporary Queen Rearing*, Dadant and Sons, Inc.

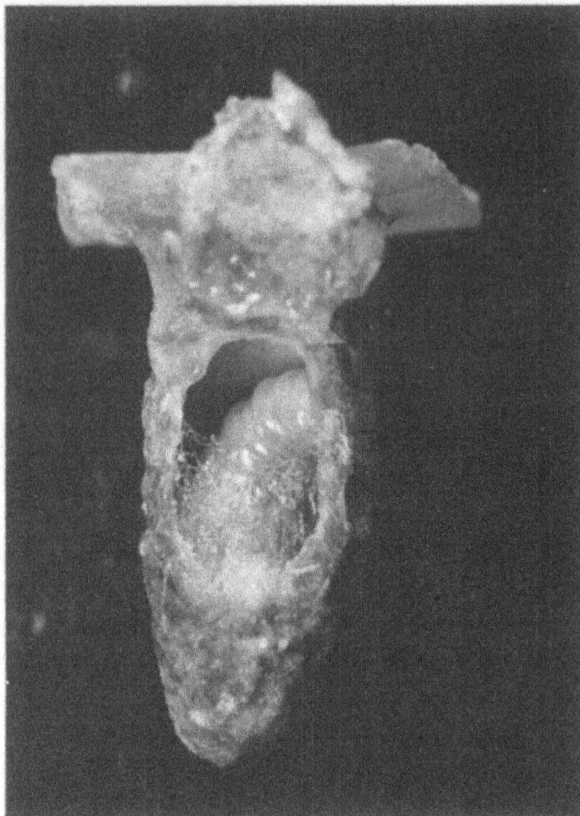

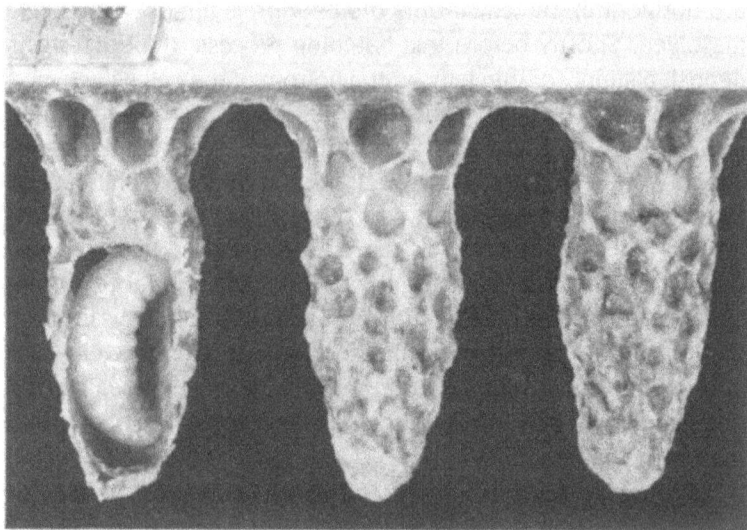

Figure 11. Prepupae. Six days from grafting.

Chapter II

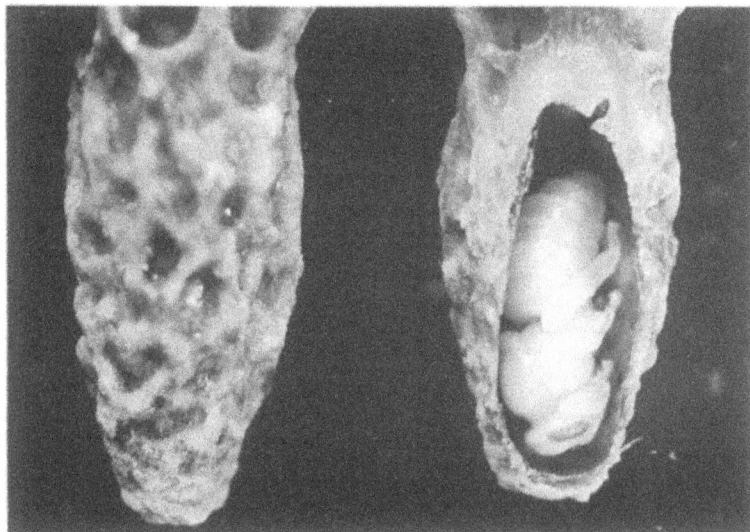

Figure 12. Pupa. Eight days from grafting. Eyes coloring. Developing queens most susceptible to damage by rough handling when in pupal stage, i.e. seven to ten days from grafting.

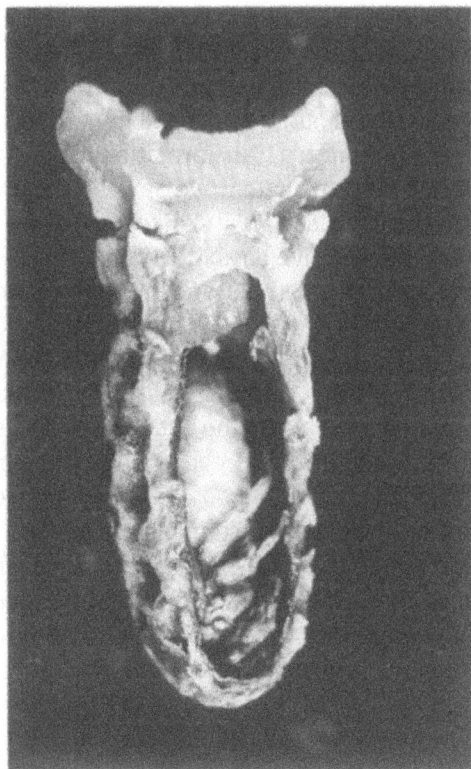

Figure 13. Imago. Eleven days from grafting.

Figure 14. Newly emerged virgin. Thirteen days from grafting. From *Queen Rearing,* Dadant and Sons, Inc.

moults, four times at approximately 24-hour intervals. Following the last larval moult, the queen larva eats ravenously and spins a *cocoon* of silk over the lower end of the sealed cell and about three-fourths up the sides of its cell (Fig. 10). The larva continues to eat, then stretches out with its head toward the sealed opening of the cell and goes into a transformation stage, the *prepupa* (Fig. 11), which lasts approximately two days. At the end of this period the prepupa casts the last larval skin to expose the immature *pupa* (Fig. 12). For the next five days the transformation from larva to adult continues both externally and internally. Some 15 or 16 days after the egg was laid the pupal skin is shed and the *imago,* or adult bee, appears (Fig. 13). Shortly thereafter the virgin queen begins to cut her way out of her cell with her mandibles. When the cap is partially cut away, she pushes it to one side and crawls out (Fig. 14). Sometimes the cap springs back in place; it may even be fastened shut with a bit of wax by the worker bees. It is not unusual for the bees to fasten this cap in place while a worker bee is cleaning the cell or feeding on the royal jelly, thus imprisoning the worker. However, the bees cut the queen cells down within a few days after the emergence of the queens, and often leave the bases as cell cups.

The Virgin Queen

When she emerges from her cell, the virgin queen is as active as the newly emerged worker. She is sometimes downy and immature looking (Fig. 14) but at other times, if emergence is delayed, she has all her color and maturity to fly or to fight. Under some conditions, such as delay in swarm departure, the bees may delay the emergence of the young queens, and during the last few hours that queens remain in their cells they often "challenge" each other by emitting shrill "piping" sounds which, at times, can be heard outside of their hives. Bees have been seen to feed queens through a small opening in their cells, and on occasion a queen may stop her work of cutting off the capping and stick her tongue through to "ask" for food from a nearby bee. A newly emerged queen may accept food from a nurse bee or take honey from an open cell.

The queen has a pronounced instinct to destroy other queens (Fig.15), and the first activity of the virgin queen is to seek out rival queens, either in or outside of their cells. She destroys numerous queen cells by cutting a small hole in their sides and inserting her long curved sting into the unfortunate occupants (Fig.16). During this task, the worker bees seem to give considerable assistance by tearing down the cells and dragging out the unemerged queen. Virgin queens pay little attention to unsealed queen cells and the bees soon discontinue their construction.

The abdomen of the newly emerged virgin queen shrinks somewhat after the first few hours, due largely to the discharge of liquid waste materials. She may thereafter be found in any part of the brood nest resting or pushing her way among the bees. She is more excitable than a laying queen and may fly from the combs or hide under a mass of bees if the colony is disturbed. The food of virgin queens appears to be mostly honey until after the queen mates. Virgins need lightness and strength for the mating flights and they derive strength from the honey.

Mating of the Queen

The virgin queen retains the navigational ability of worker honey bees and she mates with drones while flying in the open, never in the hive, and seldom, if ever, inside any enclosure. Within three to five days after she emerged from her cell, she may take one or more orientation flights that serve to mark the location of her hive,

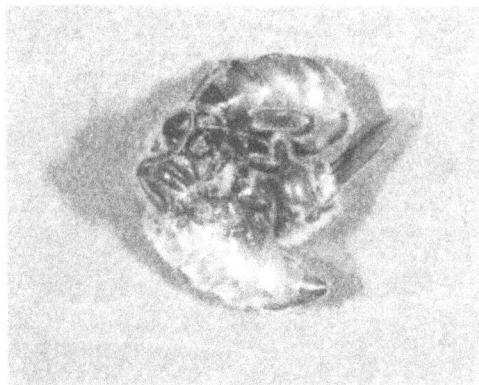

Figure 15. Virgin queens fighting.

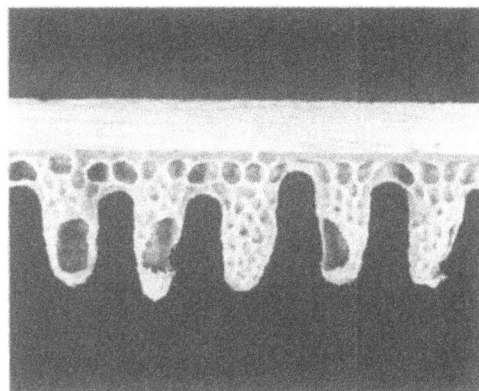

Figure 16. Queen cells destroyed by a newly emerged virgin queen.

and mating sometimes occurs on these flights. Under usual seasonal flight conditions, the queen will take her nuptial flights between the fifth and fourteenth day after emerging from her cell. If a virgin queen is confined to her hive by unfavorable weather conditions for a period of approximately three weeks without an opportunity for a mating flight, or if she is unsuccessful in her attempts to mate during this time, she may start to lay and thus become a *drone layer.* Commercial queen breeders often destroy all virgin queens that have failed to mate within fourteen days after emergence, for they believe that queens which mate later tend to be inferior layers. This assumption is supported to some extent by Woyke (1976) whose studies revealed that queens that were more than 14 days old, and instrumentally inseminated with 8 cubic mm of semen, stored only about one third the number of spermatozoa in the spermatheca as did queens 5 to 10 days old that received comparable insemination. This may not mean that queens whose mating was delayed by inclement weather are seriously impaired, because such queens can build and maintain a normal colony for a season.

Experience indicates that it may be necessary to locate a mating yard several miles from other colonies to assure racial or strain purity of mating under natural conditions. The location of mating yards on Kelley's Island and on Pele Island by the United States and Canada respectively, are excellent examples of attempts to influence matings. An established and successful breeding program in Germany relies on mating control of queens on an island where virgins to be mated, and drones for their mates, are brought to the island in appropriate colonies (Jost Dustmann, personal communication).

Stock mating control can be achieved by locating special mating apiaries in prairie expanses, in latitudes or elevations where feral bees do not survive weather extremes, or areas with sparse food or water resources. Nuclei with ripe queen cells or mature virgins are brought to the mating area along with colonies strong with mature drones of the selected drone mothers

During her mating flights the queen may fly a considerable distance from her hive, if drones are not plentiful, before she is successful. Mating flights extend from five to thirty minutes, being longer in the early spring than in the summer when the weather is warmer and drones are more numerous. Very few reports have been recorded of free flying queens being observed in the act of mating, but it is known that the sex organs of the drone are detached from its body during the act and remain attached to the queen on her return to the hive. This telltale "sign" of mating may remain attached to the body of the queen for a short time. Woyke and Ruttner (1958) have shown that the "mating sign" is the bulb of the penis, and that it separates along the surface of a preformed layer of chitin in the bulb wall allowing the fully everted endophallus to pull away undamaged from the bulb, effecting the separation of the queen and drone.

For many years it was considered that queens mate only once in their lifetime. There were, though, infrequent reports by beekeepers that one or more of their virgins mated on more than one flight. Nolan (1932) appreciated the significance this had to bee breeding, and he accumulated reports of multiple mating and presented them as a paper at the Maryland State Beekeepers Association 23rd Annual Meeting in 1932.

Roberts (1944) subsequently made a study of mating flights of 40 queens. In describing his results he observed:

> "The 40 queens made a total of 88 flights, the number ranging from one to five per queen. All queens returned once or twice with the mating sign. Nonmating flights lasted from 4 to 19 minutes, with an average of 11 minutes. Mating flights lasted from 5 to 21 minutes, averaging 14.4 minutes. Nineteen queens mated on their first flight and nine of these did not leave the hive again. Nine queens left the hive only twice and mated on both flights. Twenty-one mated only once and nineteen mated twice. Two of the single-mated queens made additional flights after once mating. Of the queens mating twice, seventeen mated on successive days and two had an interval of one day between matings".

This revelation that queens often make more than one mating flight aroused interest in whether queens mate with more than one drone on a single flight. Woyke (1955) by measuring the quantity of semen produced by individual drones, and the quantity of semen in the oviducts of queens just returned from their mating flights, found the queens mated with 8 to 10 drones on one flight. This amount of semen will fill the oviducts to near capacity. Each queen returned to her hive with her sting chamber filled with a "mating sign".

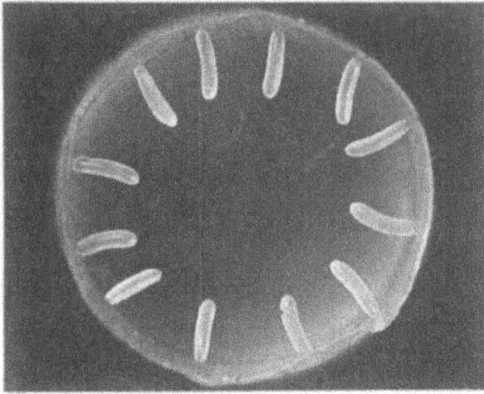

Figure 17. Adhesion of eggs by their smaller ends to a surface. In the ovaries, the oviducts and during oviposition, the small end of the egg is oriented posteriorly. From *Queen Rearing*, University of California Press.

Adams et al. (1977) estimated that at least some queens mate with as many as 17 drones. It is reasonable to question whether the several last drones to mate with the queen deposited more than a pittance of semen. Woyke's data indicates that semen from about 8 drones supplies a normal insemination.

The "mating sign" left by the drone at mating fills the queen sting chamber and would seem to preclude further matings by the queen on the same flight. Gudren Koeniger (1986), in a study of queens mating in simulated flight, observed that a succeeding copulating drone thrust the copulatory organ into the sting chamber beside the "mating sign" that was left by a preceding drone slipping the "mating sign" out as the penis everted, and then ejaculated semen into the queen reproductive tract.

Queens have seldom been seen to leave the hive after they have started to lay, except to leave with a swarm.

Oviposition

Laying of Eggs

Within two to five days after mating, the queen begins to deposit eggs in worker cells. The small end of each egg is fastened with a sticky substance (Fig.17) to the bottom of the cell. Queens which have not mated, or are laying abnormally, may lay two or more eggs in a cell and sometimes fasten them to the sides of the cell. A good queen will lay regularly in an area on one side of a comb and then lay in the same area on the opposite side, gradually extending her brood to the adjacent combs. She first inserts her head into a cell to inspect it for size (Niko Koeniger 1970), and then either passes on to another cell or inserts her abdomen into the cell and extrudes the egg so that it is attached to the bottom of the cell, which in the hive is vertical and the sides of the cells are horizontal.

Production of Eggs

The two *ovaries* of the queen (Fig.18), each composed of a number of tubules called *ovarioles*, are located in the anterior portion of her abdomen. The total number of ovarioles varied from 260 to 373 in one study made on 280 queens (Eckert 1934). The *primary* egg cells, the *oogonia* (Fig.19), from which the eggs originate, are located in the anterior end of each ovariole. The oogonia multiply several times as they progress down their ovariole toward the base of their ovary. The divisions stop and the daughter oogonia

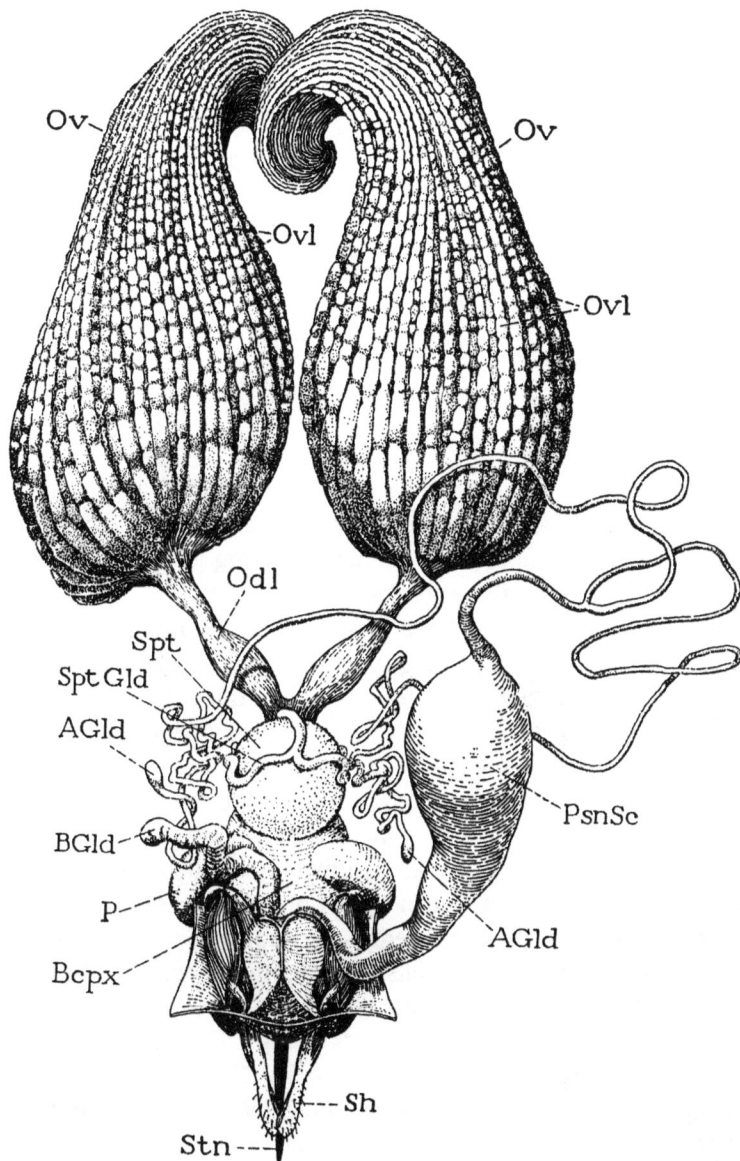

Figure 18. Ovaries and associated structures of the queen. From Snodgrass 1910. Agld - poison gland of sting; Bgld - accessory gland of sting; Bcpx - bursa copulatrix; Odl - lateral oviduct; Ov - ovary; Ovl - ovariole; P - lateral pouch of bursa copulatrix; PsnSc - poison sac; Sh - sheath of sting; Spt - spermatheca; SptGld - spermathecal gland; Stn - sting.

are now undifferentiated egg cells, known as *oocytes*. The oocytes gradually develop, and grow at the expense of accompanying *nurse* cells. When the oocyte reaches its full size and just before it passes into the oviduct, the *chorion*, which is the egg covering or outer membrane, is formed to complete the egg. A reticular opening through the chorion in the larger end of the egg, the *mi-*

Figure 19. Origin of the egg and its growth and development in the ovariole. From Snodgrass, 1925. FC1 - follicle cells; Epth - epithelium; NCL - nurse cells; Nu - cell nucleus; Ooc - oocyte; Oog - oogonial cells.

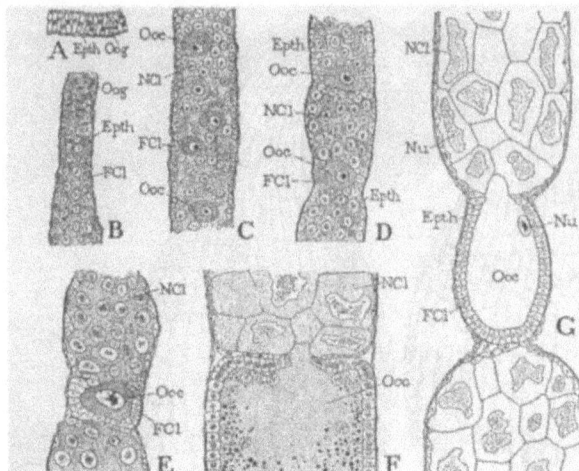

cropyle, provides for entrance of the *sperm* into the egg. Approximately two days elapse from the time the egg is started by the primary egg cell until it is completed.

A normal queen may lay an average of 1,200 to 1,500 eggs each 24 hours—about 5 to 7 eggs per ovariole. Some brood counts made by early workers indicated that some queens have laid as many as 3,000 eggs a day for a short period. The egg-laying rate depends on the strength of the colony, the amount of available space, and the amount and kind of food fed to the queen by the nurse bees. Some queens may lay somewhat irregularly at first, but after a few days a well-mated queen lays an increasing number of eggs daily until she reaches her potential. In small nuclei or in larger colonies with limited egg-laying space, a queen may lay two or more eggs in some cells after nearly all cells have eggs. This trait disappears as more empty cells become available. A well-mated queen may lay drone or worker eggs "at will" for several years, depositing them in the proper cells. The queen is able to distinguish worker cells from drone cells by measuring their diameters with her forelegs (Niko Koeniger 1970).

Fertilization of the Egg

As an egg moves down the median oviduct and into the *vagina* it passes a point where the duct from the spermatheca opens into the vagina. An invagination of the ventral vaginal wall, the *valvefold*, extends up to the spermathecal duct opening and causes the micropyle at the large end of the egg to brush against the

Chapter II

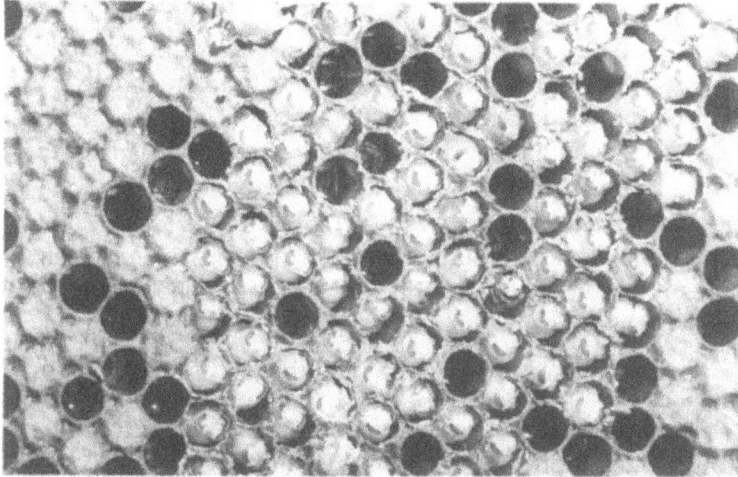

Figure 20. Worker pupae in drone cells. Eggs were laid by a mated queen that was confined to one drone comb.

opening of the duct as the egg moves into the vagina. If the egg is to be fertilized, it pauses momentarily as a number of sperm are released, suspended in spermathecal fluid, from the spermatheca into the duct. The sperm pass down the spermathecal duct and one or more sperm enter the slightly protruding egg cytoplasm and are withdrawn into the egg through the reticulated micropyle. A sperm unites with the nucleus of the egg to effect fertilization. If the egg is not to be fertilized, the valvefold is probably lowered as the egg passes through vagina, and the egg is laid unfertilized. All eggs produced in the ovaries of the queen are unfertilized and will develop into drones when laid unless they are fertilized as they pass through the vagina.

A normally mated young queen seems loath to lay drone eggs for a number of weeks after she begins to lay, and when confined to a comb containing only drone cells she may lay eggs which will produce female larvae (Fig. 20).

Queens lay fertilized or unfertilized eggs in a systematic manner and lay them in the proper cells. A few drone eggs are laid in the spring in advance of the swarming period and provide drones of proper maturity to become mates of virgin queens that follow the issuance of the first swarms

Classification and Judging of Queens

Purebred Queens
A queen that produces uniformly marked workers and drones may

be considered to have originated from a queen mother of unmixed ancestry and to have mated with drones of the same race. Since a queen mates only in flight, and with many drones, her chance of producing workers of different colors is enhanced. The sperm spread out in the spermathecal fluid after they enter the spermatheca, (sperm from one drone are not attracted to other sperm of the same drone nor to sperm of unrelated drones) but dispersion is not entirely uniform unless the sperm of the queen's mates are thoroughly mixed before they are deposited in the oviducts of the queen. This was shown by the research of Moritz (1983), Laidlaw and Page (1984), and Page, Kimsey, and Laidlaw (1984).

The mating of a queen does not influence the color of her drones since drones originate from unfertilized eggs; consequently, the color of the drones produced by a queen is a good indication of her purity of breeding, but not of her mating.

Commercial Classification of Queens

In the early days of queen rearing, when importation of European stock, especially Italian, was popular, maintaining the integrity of the imported stock was a problem. The feral bees were descendants of the dark *Apis mellifera mellifera* bees that were brought to America by early settlers. These bees, finding the environment hospitable and devoid of honey bees, gradually occupied much of the eastern and southern area of the U.S., and their drones dominated mating of virgins in their vicinity.

Beekeepers rearing and selling queens of their imported stock learned they could not guarantee their virgins would be mated to imported stock drones. This situation was alleviated, partly at least, by classifying the queens so the buyer could rely on receiving the queens that were wanted. As the feral *Apis mellifera mellifera* bees were gradually replaced by Italian stock the use of queen classification was abandoned. Nevertheless, the classification of queens by a standard system is useful and may again find a place in commercial queen rearing as Africanized bees invade commercial queen rearing areas.

Queens were classified as untested, tested, select tested, and breeders. *Untested* queens were those which are sold soon after they began to lay. If queens were held in their nuclei until their bees emerged, so that the producer could determine their purity of mating, they were said to be *tested*. *Select-tested* queens had

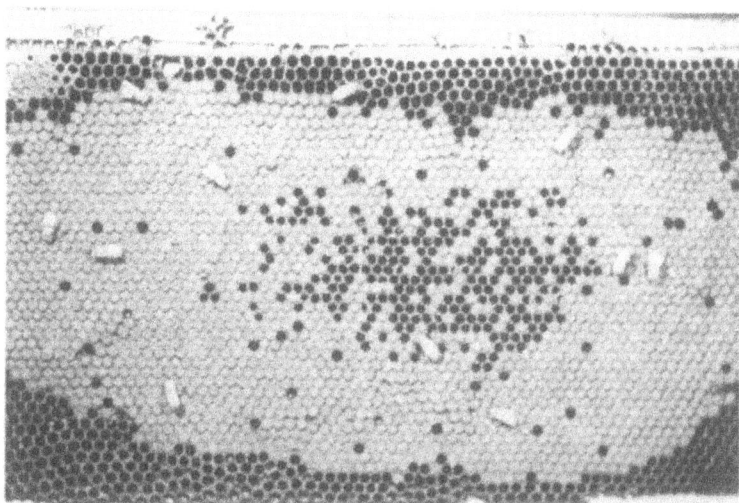

Figure 21. Emerging normal worker brood. Queen started laying in the part of the comb where bees are emerging. She extended her oviposition toward the margins of the comb with some brief temporary by-passing of empy cells. Some scattered cells were missed and did not receive eggs later, or eggs homozygous at the sex locus were laid in them and the larvae were removed by the bees.

been held in producing colonies until the producer could judge them not only for purity of mating, but also for disease resistance, productivity, gentleness, and other characteristics. Queens to be used as *breeders* were kept until it was known whether their daughter queens were capable of producing colonies with most of the characteristics desired by the buyer. This required keeping the prospective breeder queens and several of their daughters at the head of producing colonies for one or two seasons, with records of honey production, industry, gentleness, wintering qualities, swarming propensities, resistance or susceptibility to various brood and adult diseases, as well as other qualities.

Most queens sold commercially fall in the untested class, but if these are produced by a queen rearer who attempts to control the types of drones in the area in which the queens are reared a large percentage of the queens will be purely mated, or nearly so, and satisfactory. The longer the queens are held in their nuclei or in producing colonies the more costly they become to the producer.

The color and appearance of a queen attracts attention, but the true value of a queen is her attainable and maintainable oviposition rate and her longevity.

How to Judge Queens

The performances of a queen, and of her colony, are the best criteria for judging the queen's excellence. If a queen is prolific enough

Figure 22. Brood of failing queen that is nearly a complete drone layer. Eggs homoallelic at the sex locus laid by a queen mated to drones closely related to her and to drones not related to her will display a similar pattern.

to maintain a suitably strong colony, and if the colony has a majority of the characteristics desired in a producing colony, then the queen is a good one regardless of her size and looks. One should judge queens by looking at the brood and the population of the colony as well as at the queen herself. In a colony headed by a good queen the brood is concentrated in a rather condensed brood area and has very few empty cells scattered among the sealed brood (Fig. 21). The presence of drone brood in worker cells among normal worker brood is evidence of a poor or failing queen (Fig. 22). The amount of brood should be consistent with the strength of the colony, the availability of pollen and honey, and the environmental conditions of the season.

A young queen is more active than an old queen and is more likely to push under a group of bees or to crawl around on the comb and hide when the colony is disturbed. She is also brighter looking because the hairs on her thorax and abdomen have not been rubbed off. As she ages, she becomes more shiny as hair is rubbed off portions of her body. For use as a breeder, an old queen may be just as good as, or even better than, a younger queen, because age may indicate the possibility of an inherent vitality and longevity which younger queens have not had time to demonstrate.

Chapter II

THE PRODUCTION OF QUEEN CELLS

Beekeeping rests on the queen. Yet, the queen does not forage, process nectar, secrete wax, defend the colony, nor rule. The queen lays eggs that are the lifeblood of her colony whose members are rapidly replaced by others. She determines the characteristics of the colony, and because of her potential life-span of two years or more, the profile of her colony remains almost the same while numerous similar populations of workers and drones come and go. The populations are not identical because of the imperfect mixture of sperm in the spermatheca, but are nearly so and are identifiable as the same colony. The queen is the beekeepers' ally, and, as such, her own genetic composition and development are important to beekeepers.

Queens are produced in one way only—the attending bees feed the developing female larvae suitable food and in quantities sufficient to develop the characteristics of the queen caste. *Bees feed the developing larvae properly when an abundance of nurse bees are present, and when they have a good income of nectar and pollen. In the absence of these foods, substitutes such as sugar syrup and pollen supplement must be supplied.* A shallow super of honey is sometimes placed on the bottom board in lieu of feeding syrup, or in addition to feeding syrup, and the honey is moved up by the bees into the broodnest. Pollen may be fed as stored pollen in combs, as dry pollen, as pollen cakes, or as pollen dusted into combs and sprayed with syrup. Pollen is often mixed with substitutes and yeast.

Determination and Development of the Queen Caste

A queen begins her existence within a fertilized egg. An embryo begins to form about the time the egg is laid, and in three days embryonic development is finished with the formation of a minute larva. The egg hatches in three days and a small amount of food is provided in the cell with the egg at the time of hatching. Food is then supplied until the cell is sealed. The embryo that forms within an oviposited fertilized egg is female (except when the sex alleles are the same—see Chapter VII) but it is, as yet, neither worker nor queen. The caste of the larva that hatches from the egg is determined during the third day following hatching. Larvae that hatch from eggs deposited in queen cells, or which were otherwise selected very early by the bees to become queens, receive an abundance of royal jelly during the first two days following hatching and until the cells are sealed. Larvae hatched from eggs laid in worker cells and destined to become workers receive worker jelly, but a lesser amount of food than queen larvae the first two days following hatching from the eggs and also until the cell is sealed. Otherwise, the food appears to be much the same during this first two-day period.

This period is critical for a presumptive queen larva, and a scarcity of food might prevent the resulting queen from reaching full physiological development, though she may appear to be a normal queen. That larvae which are hatched in queen cells normally receive a greater abundance of food during the first two days than even well-fed worker larvae does not necessarily mean, however, that they are actually better fed during this period. There is a limit to the amount of food a larva can consume. But it is likely that larvae hatched in worker cells and fed *sparingly*, as judged by the amount of brood food in the cell, are underfed and will produce queens inferior to those adequately fed from the time of hatching. Inferiority may be expressed in several ways—including size as a physical attribute, oviposition rate, or longevity.

The bee-gathered food of bees comes from three sources—nectar, pollen, and water. From these main materials, the worker bees elaborate the food they need for themselves and for their queens, drones, and developing brood. Nectar provides the carbohydrates, some minerals, a portion of the vitamins, and some of the water needed for the production of energy and for the building of various elements of the cells of the body. Pollen provides the only natural source of protein for the bees, and it also contains vitamins, minerals, and lipids. Water dilutes honey, provides some minerals, and serves to maintain a balance of water in the blood and body tissues. Adult bees can maintain themselves with

Chapter III

honey or sugar syrup alone, but require pollen, or a suitable substitute, for the production of brood food.

Brood food originates from the digestion of honey and pollen in the digestive tract of the worker bee. The materials are broken down into simpler compounds by various *enzymes*, and are absorbed into the blood. A pair of glands located in the head region of worker bees, known as *pharyngeal*, or brood-food glands, take from the blood the constituents needed to produce a secretion which is passed down through their ducts to the mouth cavity near the base of the tongue. This secretion, together with secretion of the *mandibular* glands, the addition of suitable amounts of sugar, and possibly secretions from other glands as well, is known as *worker jelly* if fed to worker larvae, or as *royal jelly* if proportions of its constituents are modified and it provisions queen cells. As it appears in the queen cells, the secretion is milky white in color and has the consistency of thick cream. If removed from the cells and kept at room temperature it turns slightly yellowish. Royal jelly kept in a deep freeze at 0°F shows little visible deterioration for periods up to several years (Smith 1959).

The composition of queen larval food differs from the composition of worker larval food. One of the most important differences appears to be higher sugar content which seems to entice a queen larva to eat ravenously, which may in turn stimulate the *corpora allata* of the queen larva to secrete a high level of *juvenile hormone (JH)* that is believed to be intrinsically involved in female caste determination (see Herbert 1992 and Beetsma 1979). Although the queen caste is irreversibly determined by the end of the third day following hatching of the egg, the full development of queen caste features is dependent upon continuous and lavish supply of royal jelly to which the young larva adheres as it is suspended in the queen cell.

Chemical Composition of Royal Jelly

Various investigators (Herbert 1992) have found by chemical and biological analyses that royal jelly is a complex substance that varies, in some respects, according to the age and type of larva to which it is fed. It contains abundant tissue-building proteins and energy-producing carbohydrates, and also lipids, vitamins, moisture, 10-hydroxydecanoic acid, and other materials. Planta (Eckert

1934), one of the first to investigate the chemical nature of royal jelly, thought that the bees added honey and pollen after the fourth day, but more recent investigations tend to show that royal jelly composition remains rather constant, particularly so when taken only from queen cells. Pollen grains found in royal jelly from queen cells may be accidental inclusions. It seems probable, also, that the differences in the chemical constituents of royal jelly, as determined by different investigators, may have been due, at least in part, to variations in the chemical composition of the pollen and honey on which the bees were feeding.

Haydak (1943) made a comprehensive review of investigations of the chemical composition of the larval food of bees, and Haydak and Vivino (1950) extended the research . References cited in their papers constitute a valuable source of information on royal jelly and larval nutrition.

Age at Which Bees Produce Royal Jelly

The ability to produce royal jelly varies with the physiological age of the worker bees. Under normal conditions young workers are capable of producing a maximum amount of royal jelly between the fifth and fifteenth days after they have emerged from their cells. The pharyngeal glands of younger bees are not fully developed, and they deteriorate in older bees. However, the "age" of the bee depends more on the bee's physical development and physiological condition than on temporal age alone. Bees just emerging from a winter period of several months, for example, can still produce sufficient royal jelly to rear queens and worker brood. A colony made up only of old workers can still rear some brood but the resulting individuals frequently are undersized, a further indication that proper and abundant food is critical in queen rearing.

Characteristics of Suitable Cell Building Environment

To produce good queen cells, the conditions that exist in nature when a strong colony produces cells under the swarming impulse should be approximated. At such times the colony is usually at its peak of strength, thus ensuring an abundance of food, nurse bees, wax secretors, ventilators, and foragers. The combs in the hive are well filled with honey, pollen, and brood. Incoming nectar and pollen create a condition favorable to the production of both wax and brood food. The crowded condition of the brood nest and the restriction in the available space to rear brood result in an over-abundance of nurse bees and an excess production of royal jelly. Comb builders are stimulated to wax production by the processing of nectar and the storage of honey. Drones are being produced, and mature drones are present in large numbers. This combination of factors generally results in the preparation of numerous queen cell cups in which the queen will lay, in the construction of queen cells, and, finally, in swarming unless unfavorable weather intervenes.

These environmental conditions are not always present when supersedure queens are reared or when queens are reared under the emergency of queenlessness. The main factor in favor of supersedure queens is the small number of cells which are usually produced.

If one wishes to rear queens at a more convenient time than during the swarming season, it is necessary to simulate, or to improve on, natural factors which are present when cells are built naturally under optimal conditions.

A colony of bees can be strengthened by the addition of young bees and emerging brood from other colonies. When there is an abundance of pollen in the combs or there is a good pollen flow, the production of brood food can be increased by feeding a light (one part sugar/one part water by volume) sugar syrup. Feeding

Figure 23. Supplying pollen to a colony by means of a pollen-soybean cake. The cake must be placed immediately above the brood area.

Figure 24. Syrup feeding through a hole in the cover. Syrup is in a jar or pail.

Figure 25. Feeding syrup by division board feeder within the hive (left), by Boardman entrance feeder (center, and through a hole in the cover (right).

a pollen supplement (Fig. 23) along with sugar syrup also stimulates brood rearing and an increase of brood food if unsealed brood is present.

Stimulative feeding started in early spring will cause a queen to increase her egg laying and produce drones earlier than usual, and the colony will be ready to rear queen cells sooner than it would otherwise. The same stimulative methods will encourage a queenless colony to produce more brood food if young larvae are added to the colony. Both queenright and queenless colonies

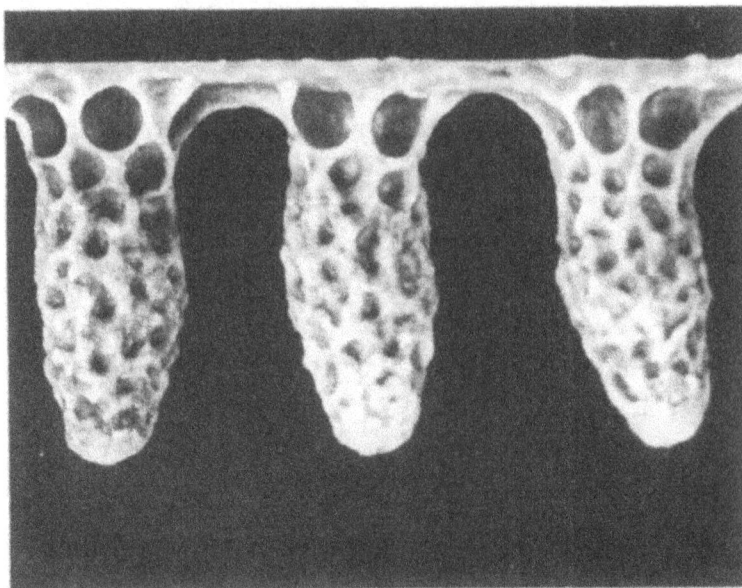

Figure 26. Well formed sealed queen cells.

Chapter III

build good queen cells as long as the relation between food and the colony population of bees of suitable age is maintained.

It is obvious that sugar syrup should be fed continuously to cell-building colonies (Figs. 24, 25) to simulate a constant nectar flow, and to provide incoming sugar during inclement weather. Sealed queen cells should be large, well formed, and sculptured (Fig. 26), and are more likely to have this sculpturing when there is adequate income of nectar or syrup.

The processing of nectar or sugar syrup stimulates the production of wax. Beeswax is elaborated from carbohydrates by the wax glands of worker bees. These glands reach their greatest productivity during the second and third weeks after the bee emerges from its cell. Older bees can produce wax, but less abundantly than this younger age group.

Nectar processing bees consume part of the sugar syrup, or nectar, while they are inverting sucrose to dextrose and levulose and evaporating the excess moisture, and they involuntarily produce wax. Too much crowding, however, can lead to excessive burr comb and to enclosure of the cells in a web of comb. A frame of foundation put into the broodnest will lessen burring of the queen cells, and also the probability that the colony will swarm.

The number of satisfactory queen cells a colony will build from one graft is related to the size of the cell-building bee population in the cell building colony. Excessive crowding and inadequate ventilation will cause many of the bees to leave the brood nest. This results in production of poorer cells. Additional ventilation and suitable clustering space for the field bees can be provided by increasing the depth of the space beneath the bottom bars of the lower brood chamber with a two-inch rim. Or better yet, a shallow body of dark combs *below* the lower brood body (Fig. 27) provides space for foragers at night or during inclement weather and avoids overcrowding the broodnest by forager bees that have few, if any, in-hive duties (Fig. 28). This, with ample space *above* the broodnest to store honey, has proved to be an effective deterrent to swarming. It has the added utility of protecting the lower edges of brood combs from winds and cold so that brood will extend to the frame bottom bars.

If the shallow body is fastened to the bottom board, the hive can then be moved with the bottom board securely fastened to the hive; and brood bodies above the shallow body can be rotated

Figure 27. "Shallow" body of combs on bottom board provides space for forager bees. The entrance is fitted with a Boardman feeder and robber screen. From Laidlaw, *Contemporary Queen Rearing.*

Marked Foragers Found in Hive on the Evening of September 7, 1994

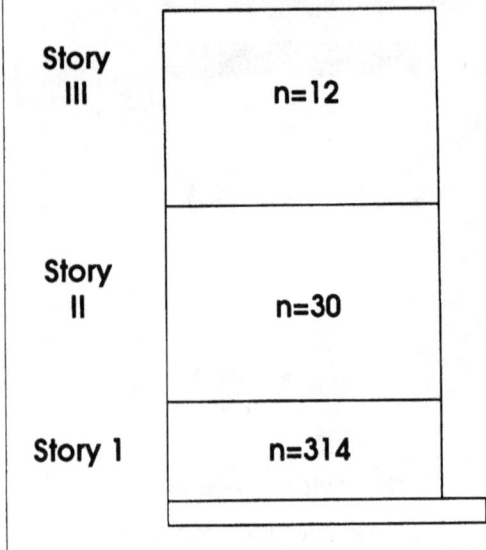

Story III	n=12
Story II	n=30
Story 1	n=314

Figure 28. Marked foragers found in a hive on the evening of September 7, 1994. This shows the stratification of worker bees of a colony in the hive. Returning foragers were screened from the hive entrance, marked and allowed to enter the hive. When flight had ceased the marked bees were located and counted. Their locations in the hive verified that foragers clustered below the brood area and contributed little to brood nest crowding if room below the brood is available. Bees tend to occupy their area of activity in a hive when space there is sufficient. Experiment performed and drawing made by Kim Fondrk.

as they normally would be. A shallow division board feeder against each side wall of the shallow body would provide permanent feeders to be used as needed.

The production of royal jelly requires the consumption of large amounts of pollen by nurse bees. Combs of pollen are placed near the cells that are being built. Combs of pollen are usually available from other colonies. If these are lacking, pollen supplements should be used.

Pollen Supplements

Feeding pollen supplements as a means of stimulating brood rearing during periods of pollen shortage, or to increase colony strength for specific purposes is a common practice. The most commonly used supplement to natural pollen is expeller- processed, soybean flour with added pollen or yeast. Expeller- processed soybean flour may not be available, and fully toasted soybean flour (Erickson and Herbert 1980) with added powdered yeast may be used. Pollen in the proportion of 1 part of pollen to 10 parts of soy flour increases the palatability of the flour to the bees. These supplements are frequently fed dry by mixing the flour and bakers' yeast (*Sacharomyces cerevisae*), brewer's yeast (*Saccharomyces fragilis*) or Torula® dried yeast (*Candida utilis*) Type S in the proportions of nine parts of flour to one part of yeast, and placing the mixture in the apiary in a box with a lid sufficiently ajar to admit bees but to keep out wind and rain. This mixture, either dry or in cake form, can be fed conveniently on a feeder screen on top of the brood bodies (Fig. 29). A cake made by mixing sugar syrup, or high fructose corn syrup, with the yeast and soybean flour so that the dough is firm enough to hold its shape, can be placed on the top bars immediately above the brood area and covered with waxed paper (Fig. 23) or fed on a feeder screen (Fig. 29). Convenient, commercially prepared pollen supplements are readily available from bee supply firms.

Pollen traps are sometimes used on populous colonies to collect pollen pellets from the legs of bees during heavy pollen production. This pollen is then stored in a freezer as is, or is dried and stored in an airtight container (so that wax moths and various other insects will not feed on it) until it is needed for stimulative feeding. The dried or frozen pollen pellets are moistened with

Chapter III

water and then mixed with soybean flour, yeast, and sufficient 50/50 sugar syrup to make a soft but firm cake. The addition of terramycin will aid in preventing AFB. The terramycin is dissolved in the water used to make the syrup, using one and one-half teaspoons of TM-25 to each pound of sugar. The bees are supplied with the pollen cakes as long as they do not have access to a plentiful supply of fresh pollen and as long as increased or sustained brood rearing is needed.

Many queen breeders prefer to collect combs of pollen from producing colonies for their cell builders. There is no substitute for natural pollen gathered by the bees or stored by them in their combs.

Selection of Breeding Queens

Queen rearing methods used by beekeepers who produce queens in small numbers and those used by beekeepers who produce queens on a commercial scale are fundamentally the same, though the manipulative practices and equipment may differ. Each must select breeder queens to maintain or improve the stock. Each must obtain and prepare larvae for the cell builders and prepare cell-building colonies to provide the proper environment for the developing larvae, and each must provide nuclei to care for the queens while they mature sexually, mate, and begin to lay.

The selection of breeding queens is the first important step in queen rearing. However competent the rearing methods, the resulting queens will be inferior if the stock is poor. Moreover, to maintain the quality of the stock the selection of drone mothers is as important as the selection of queen mothers. By producing queen cells for general requeening from the purely mated and high-producing colonies, which also have most of the other traits one wishes to perpetuate, mothers of the potential mates of supersedure queens are automatically selected (Laidlaw 1956). It is important to rear virgins from several breeder queens to reduce undesirable inbreeding.

Figure 29. Pollen and pollen supplement feeding screen. A robber screen is on the entrance.

Producing Queens under Non-Controlled Conditions

Naturally built supersedure and swarm queen cells from colonies of good stock will often yield excellent queens, and when few cells are needed they are a valuable source of new queens. The sealed cells are cut from the combs and given to dequeened colonies, or are put into nuclei for the virgins to mate for later use.

Swarm Cells

During the swarming season, many colonies preparing to swarm build more cells than are needed to assure the survival of the colony. Cells of several ages are found in such colonies; a few cells may be started each day for a period as long as 7 days. The preconstructed cells are likely to be the largest, although cells built around very young larvae in worker comb cells and well cared for may be as large when finished. Swarm cells, as a rule, yield excellently developed queens.

Supersedure Cells

Colonies will supersede their failing queens unless the beekeeper replaces the queens first. When queens are "purely" mated and their colonies have no strikingly bad traits, supersedure cells can be used to advantage. The removal of supersedure cells that are nearly ready to emerge frequently results in the colony's building others, and this reaction may continue as long as the queen is able to lay and the colony is strong enough to build cells. If the colony is strong at the beginning of cell building, or emerging brood is added periodically, and if the nectar and pollen conditions remain favorable, "ripe" (nearly ready to emerge) queen cells can be removed every few days. However, because supersedure cells are built when the queen is failing, and because queen failure may occur anytime in early spring, midsummer, or late fall, this method of securing cells is not dependable. It lends itself to the production of only a few queens at uncertain times of the year. Remember that the uncritical use of supersedure cells is subject to the same dangers of stock deterioration inherent in the use of swarm cells.

Cells Produced under Queenless or Emergency Conditions

Colonies will produce queen cells when the bees are separated from their queen. This reaction may occur under a number of circumstances, among which the most common are—

1. When the queen dies, is killed, or is removed from the hive;
2. When the brood nest is divided by means of a queen excluder and the queenless division has young brood, as when a colony is "Demareed" for swarm control;
3. When a colony is divided and the queenless portion is placed in a separate location;
4. When bees are shaken into a hive, nucleus, or ventilated box without a queen, and are given combs of pollen, and honey or sugar syrup, and are supplied with young larvae.

The beekeeper may take advantage of this replacement tendency by manipulating these conditions to get a colony to build as many cells as are wanted. In each case, conditions favorable to the production of good queen cells must be present.

One of the earliest methods to produce queens was simply to remove the queen from a populous colony when nectar and pollen were plentiful and then, ten to eleven days later, cut the sealed queen cells from the combs and transfer them to queen mating hives. This method was improved later by removing all unsealed brood from such a colony and giving it a comb containing eggs and young larvae from a more desirable strain. The colony will rear a number of cells if it has access to combs of pollen and is fed sugar syrup while the cells are being built.

A colony that is starting to build queen cell cups and is making other preparations to swarm will continue to build cells if the queen is confined to the brood chamber and most of the combs of brood are placed above the queen excluder. Queen cells will be started, typically, in the queenless portion. Continuous feeding of pollen and sugar syrup is essential to assure optimum conditions for the production of the queen cells. Since the colony is queenright, it is likely to start swarm cells in the lower chamber as well as in the one above, and those in the lower chamber must be torn down every week, or moved above the excluder to prevent the colony from swarming when these cells are sealed. It is good practice to clip the queen's wings so that if the bees do swarm the queen will be lost but the bees will return to the hive. A colony in this condition will continue to rear good queen cells as long as the queenless portion is supplied every three or four days with young larvae. The queen below will maintain the strength of the colony unless she reduces her brood rearing in preparation for swarming. To maintain an adequate number of nurse bees in the queen cell building compartment, combs containing young larvae are moved up from the lower brood chamber every time examinations are made for queen cells.

Producing Cells under Controlled Conditions

When cells are produced under controlled conditions the breeding queens and the cell builders are housed in separate hives.

Obtaining and Preparing Larvae for the Cell Builders

After the breeding queens have been selected, larvae are obtained from them and prepared for the cell builders. The way this is done will be determined by whether the larvae are to remain in the cells in which they hatched or are to be transferred to queen cell cups for the remainder of their growth and development. The beekeeper with few colonies and need for few queens may prefer not to transfer the larvae, although transferring the larvae is easier.

• *The Miller Method.* The method popularized by Dr. C.C. Miller (1912) is the simplest of all the procedures and is probably one of the best for the amateur beekeeper to use. Two or more strips of foundation two or three inches wide at the top and tapering to a point within an inch or two of the bottom bar are fastened to the top bar of an empty frame (Fig. 30). The frame is then put into a breeder colony. Even if the bees are bringing in nectar, the colony should be fed 50/50 sugar syrup. To prevent the bees from building drone cells, combs of honey or sealed brood are substituted for all but two frames of brood of the breeder colony, between

Figure 30. A frame prepared with strips of comb foundation for use in the Miller method.

Chapter III

which the prepared frames are placed. In about a week the foundation will be drawn out and the comb will contain brood, with the youngest brood and eggs toward the edges.

Because the combs built from these foundation strips are not wired into the frame and are constructed of new white wax, they fall out of the frame easily if they are shaken or if the frame is turned on its side. They must be handled carefully. When the comb is removed from the breeder colony, the bees should be brushed off *gently* and the comb laid flat on a table or board for trimming. The egg-containing margins of the comb are trimmed away with a warm, sharp knife, leaving the youngest larvae in worker cells at the new comb margins. The comb is now ready to be put into a queenless cell-building colony. Cells will be built along the cut margins of the new combs.

• **The Alley Method.** The method developed by Henry Alley (1883) resembles Miller's in that the larvae are allowed to remain in the strips of cells of worker comb in which they hatched. With Alley's method, an empty *new* comb is placed into the brood nest of a breeder colony; four days later it usually contains eggs and newly hatched larvae. The comb is then cut from the frame and into strips with a warm, sharp knife by running the knife through the middle of alternate rows of cells, leaving a center row intact. With the warm knife, the cell walls of one side of the comb are shaved down to within one-fourth inch of the midrib and every second and third larva in the row of intact cells is destroyed. These strips are attached to a dark comb from which the lower two-thirds has been cut out and removed. The strip is fastened by dipping its uncut cells into melted beeswax just at its melting point, and by pressing it against the lower edge of the comb and holding it there until the wax has cooled somewhat. The cells containing the larvae are thus situated so they open downward. As soon as the strip is fastened to the comb, the frame is placed in the cell builder.

• **The Smith Method.** Smith (1949) modified the Alley Method and adapted it to large-scale queen production. Only new combs are used because the bees find them easier than older comb to remodel into queen cells. Each breeder queen is housed in a specially modified hive body which has a partition dividing it into a narrow compartment capable of taking three standard frames, and a wide compartment capable of holding six frames. The par-

Figure 31. Smith "wooden frame" with newly drawn comb for use in the breeder hive.

tition extends to within an inch of the bottom board; the space between the partitions and the bottom board is filled with a strip of queen excluder. In Smith's hive, the partition extends three-quarter inch above the top of the hive and each compartment is closed on top with a wooden inner cover. A telescope cover fits over all. A variation, whereby the partition is flush with the top of the hive and water-resistant canvas to cover the frames of each compartment is attached to the upper edge of the partition, is equally effective and is easier to work with. In both types, the small compartment has no entrance, and the entrance to the large compartment is located halfway up the side, beneath the handhold. A hole bored in the hive wall provides access to a feeder attached to the outer wall of the small compartment.

The queen is confined to the narrow compartment on three frames. Two of these, each of which remains against a side wall, have a piece of comb nine and one-half inches long by five and one-half inches wide attached to the top bar at the middle of the frame; the remainder of the frame is filled with wood (Fig. 31). The third frame contains a piece of white comb of similar size that coincides with the combs of the side frames. This frame is placed between the two side frames. The restricted comb area in the small compartment will cause the queen to lay promptly in the white comb, which can then be cut into strips and attached to bars.

Chapter III

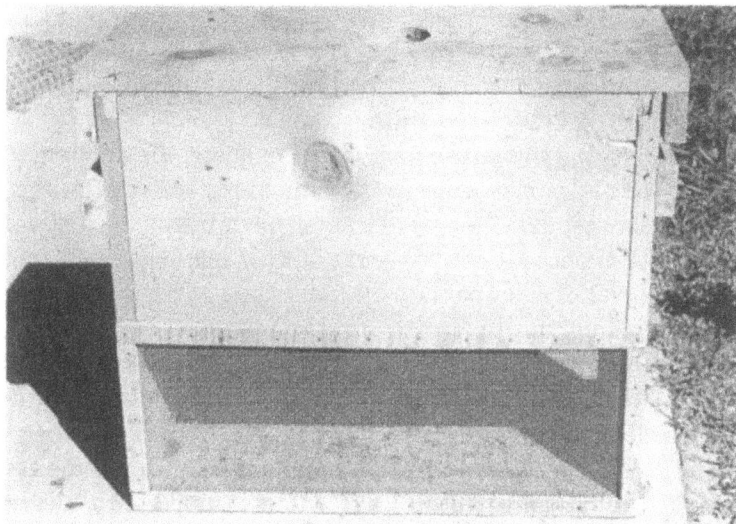

Figure 32. Swarm Box.

The white comb is obtained by fastening a piece of foundation, nine and one-half by five and one-half inches, to the top bar of an empty frame and then putting the prepared frame above an excluder in a strong colony or in the larger compartment of the breeder hive. When the cells are partly drawn the comb is ready for the breeder queen.

To make up the breeder colony, the breeding queen and one frame of brood and bees are put into the narrow compartment between the frames containing the small combs. The remaining brood and bees of the breeder colony are put into the wider compartment. The colony is fed; feeding must be continuous as long as cells are being raised. After the queen has filled the small side combs with brood, the center brood comb is removed and one of the new white combs is put in its place. Twenty-four hours later, the new comb will be well filled with eggs. It is then placed in either the large compartment of the breeder hive or in a cell finisher for incubation and for the first feeding of the larvae. This sequence is repeated daily as long as larvae are needed. By the time the fourth comb of eggs is ready for incubation, the first will have abundantly-fed larvae and is ready to be prepared for the cell builder.

The breeder colony cannot maintain itself when the young brood is removed repeatedly. Combs of emerging brood should be given to the colony at least once a week. Smith opens the starter

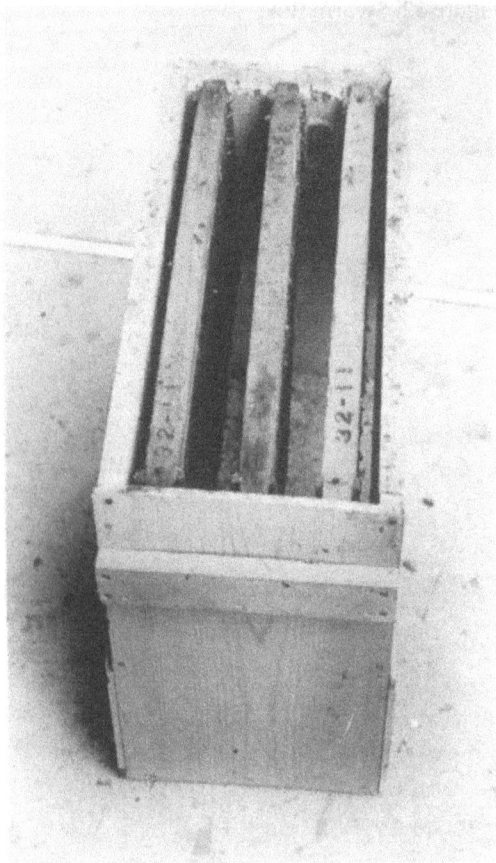

Figure 33. Swarm box interior. The side combs of honey can be replaced by division board feeders.

colony in front of the breeder hive when the cells are removed from the starter. Many of the younger bees then join the breeder colony. Young bees from one comb of the starter may also be shaken into the breeder colony as needed.

In preparing the larvae for the cell builders, the comb containing the day-old larvae is cut from the frame and into strips as in the Alley method. Then, with a small paint brush, melted wax is painted on one side of a bar and a strip of cells is pressed onto the wax. Wax is then painted on both sides of the cell strip to attach it more firmly to the bar.

At this point, the worker cells are too close together to permit the construction of queen cells from each one, so, as with the Alley method, only every third larva is left, the two between being destroyed. Two bars of the prepared cells are put into a frame and two such frames are put into a starter colony. Smith used a "swarm box" to get the cells started (Figs. 32, 33).

The small danger of overheating the larvae in the cells when the pieces of comb are fastened to the cell bars with melted beeswax can be avoided by using a split cell bar and fastening a comb strip of larvae to it by squeezing the cell walls of the side opposite the prepared side between the two pieces of the bar. Additional support can be gotten by using small brads in one piece of the bar and pushing them into the other half. If the call bars have been coated with wax before they are pressed tightly together with the strip of comb in between, the comb in between will support the weight of the bees until the cell builders can further fasten the strip in place. This technique speeds up the operation and reduces the risk of injuring the young larvae through exposure to drying. But unless the pieces of comb are held securely in place the bees will pull them down or cut them away from the bar.

Other methods of preparing the larvae for the cell builders are devised from time to time but none are as easy or as efficient as the grafting method.

• **_The Doolittle or Grafting Method._** The essential feature of the Doolittle method of producing queen cells is the transference of young larvae in worker cells to prepared queen cell cups. The larvae are taken from brood combs previously placed in breeder colonies. In this method of preparing larvae for the cell builders, the kind of comb used is unimportant as long as it is worker comb. The care of the larvae before they are transferred is just as important, however, as the care of the larvae prepared by the above methods for the cell builders. The larvae are transferred to queen cell cups when they are 12 to 36 hours old, and it is important that they be well fed from the time of hatching until they are transferred and, of course, afterward.

The Doolittle or Grafting Method of Cell Building

Obtaining Larvae

To obtain larvae for grafting, a dark comb that has been polished and conditioned above an excludered brood nest of a populous colony, is put into the brood nest of a breeder colony. The next day the comb has eggs and it is transferred into a queenright incubator-feeder colony in a body excludered from the queen but with young larvae and crowded with nurse bees. Four days later the oldest larvae are about 24 hours old and ready for grafting. If queens are being reared successively, this sequence should be maintained, preferably among different breeder colonies, so that there will be a regular supply of larvae the right age for grafting. Larvae cared for in incubator-feeder colonies from hatching to grafting are as well-fed as are larvae of comparable age prepared by the Miller or Alley methods. Grafted larvae receive immediate care in the starter or starter-finisher colony.

To ensure that the queen will lay the necessary number of eggs in the comb provided, the brood chamber of the breeder colonies is divided into three parts by means of queen excluders (Fig. 34). The queen is confined to the middle compartment, which may have one to three combs. The two side divisions are stocked with emerging and sealed brood to maintain colony strength and to produce bees to incubate the eggs. The comb in the center is exchanged with a side division comb daily as the center comb of newly laid eggs is transferred to a side division for incubation and after hatching, for larval feeding.

An efficient hive insert for assuring a continuous supply of well-fed larvae of the right age for grafting was contrived by Mel Pritchard (1932). This device has two compartments as a unit. The breeder queen is confined to three small frames of combs in a compartment that has a solid bottom, is excludered on the sides, and can be closed on top, This compartment is part of a rack that is suspended in the middle of a breeder queen's broodnest. A sec-

Figure 35a. Pritchard breeding queen insert.

Figure 35b. Pritchard breeding and queen insert in a hive.

Chapter III

ond, but non-excludered, compartment is adjacent on the rack, and it also has three small frames of combs (Figs. 35a, b).

The side combs of the excludered compartment have honey or stored syrup and preferably some pollen. The middle comb has empty cells available for eggs. Each day the comb of eggs is transferred from the excludered compartment to the non-excludered compartment in exchange for an empty comb. On the fourth day the larvae are ready for grafting (Fig. 36).

After grafting, the larvae remaining in the comb are spraywashed from the comb with water. The water is shaken out, and the comb is again inserted into the middle space of the compartment with the queen, after the current middle frame with eggs is transferred to the non-excludered compartment.

The breeder colony cannot maintain itself, and periodic additions of emerging or sealed brood on each side of the insert are necessary.

The Queen Cell Cups

Queen cell cups to receive the larvae may be obtained by the beekeeper in several ways. When few queens are needed, a supply of

Figure 34. A variation of the three compartment body has only one comb in the middle in which the breeder queen is confined.

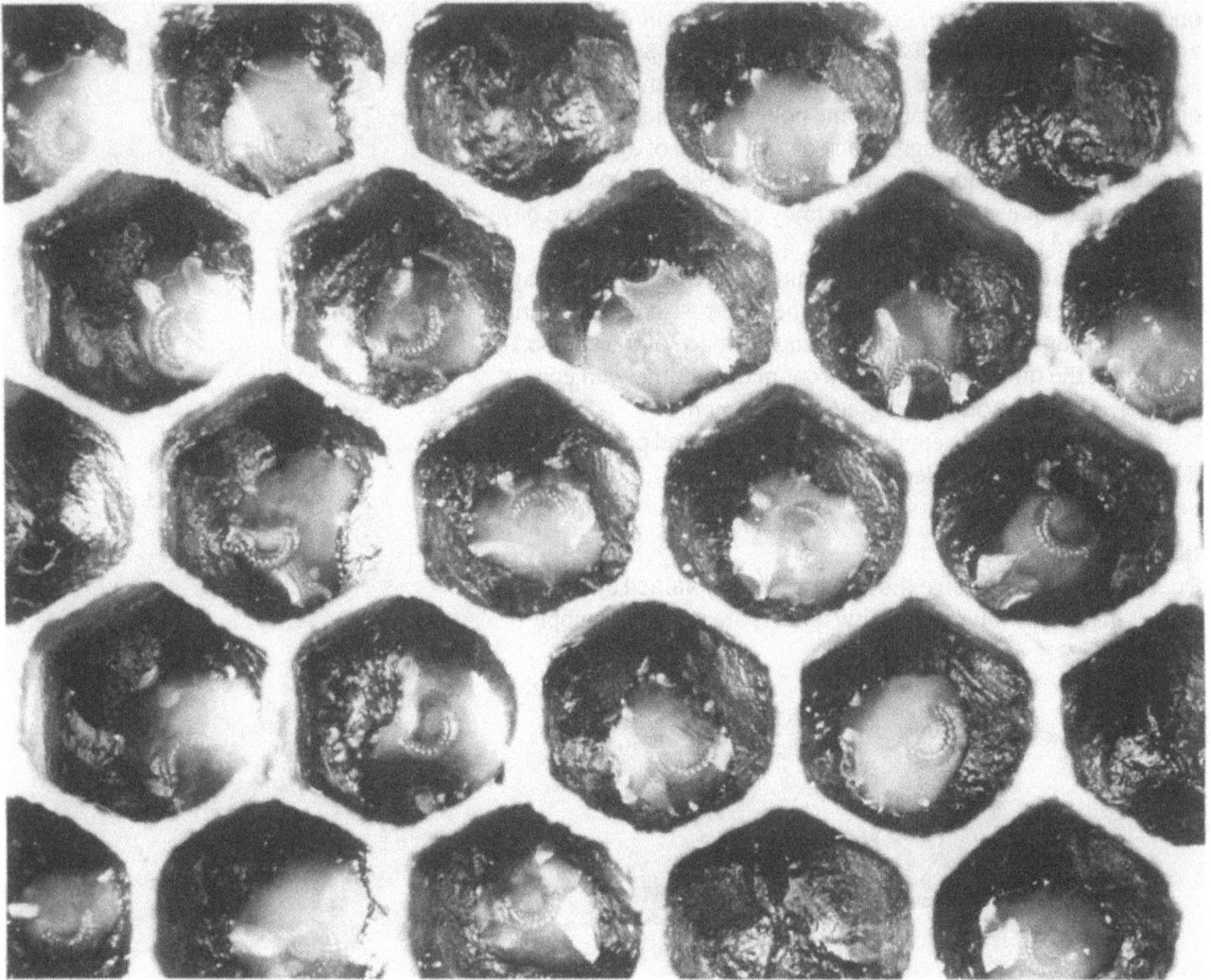

Figure 36. Larvae the proper age for grafting—about one day old. The larvae are well fed from the time of hatching.

cups can be gotten by cutting off the "embryo" cups which are present in most colonies. Doolittle followed this practice before he began making cups by dipping. Strips of drone comb that have been shaved down to within one-quarter inch of the midrib, and in which the two cells between every third cell has been mashed down, make good cups. Pressed wax cups can be purchased from supply dealers at a reasonable cost, or the beekeeper can make his own cups by dipping or by other methods. Plastic cell cups, such as the JZˢ BZˢ plastic cups, have gained popularity and are comparable to wax cups in acceptance by the bees.

Chapter III

• **_Dipping Single Cell Cups._** The hobby beekeeper usually finds a single forming stick (Fig. 37) adequate for dipping cell cups. The stick, about three inches long, is made of 3/8 inch hardwood rod. It has a three-eighth-inch diameter at a point one-half to three fourths inch from the tip, and tapers to between one-quarter and five-sixteenths of an inch at the tip. The end is rounded and smoothed to give the bottom of the wax cell cup a smooth, concave form. The forming stick can be made by the beekeeper or purchased from supply dealers. It can be tested for size by fitting the end into a naturally built queen cell cup.

The wax to be used in making cell cups should be clean beeswax. New light wax is preferable, but darker wax is satisfactory. The wax can be melted in a water-jacketed tray, or electric pan, and is kept at a temperature just above its melting point. Beeswax is flammable and care must be exercised in heating it.

When the wax is ready, the forming stick is dipped into cold water. The excess water is shaken off so the cell cup bottom will be smooth, or the stick is touched to a towel to remove the drop of water that hangs from the end of the stick. The stick is dipped into the wax to a depth of about five-sixteenths inch and is quickly removed and held in the air a moment until the wax has solidified somewhat. Then it is again dipped into the wax and quickly removed. This process is repeated four or five times. After the final

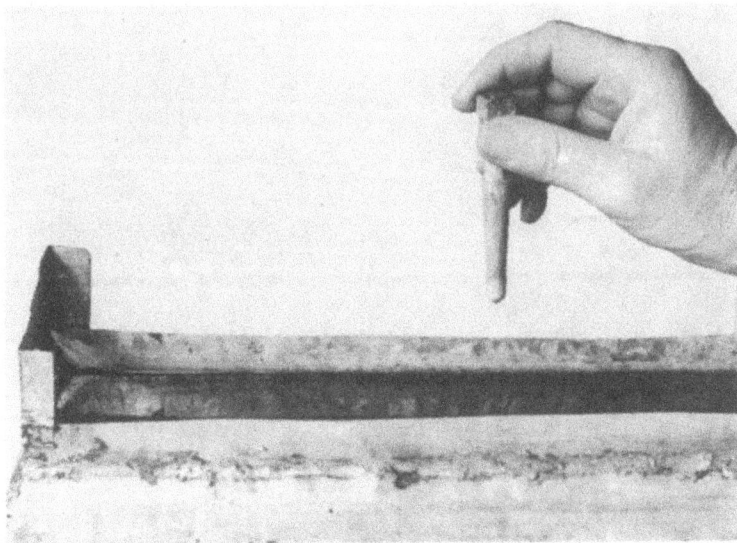

Figure 37. Dipping cell cups with a single cell mold.

dip, the mold with its formed cell cup is immersed into cold water. A gentle twist will remove the cell cup from the stick. The stick must be dipped in water again before the next cell is formed.

The prepared wax cups are attached to bars that are just long enough to fit between the end bars of a standard frame. The bars can be made from regular frame bottom bars. It is preferable that the frame to receive the bars is no wider than the cell bars. Blocks one-quarter inch thick, three-quarters inch wide, and one and one-half inches long are nailed to the inside of the end bars to form ledges on which the cell bars can rest. The lower blocks touch the bottom bar. The lower ends of the second pair of blocks are spaced about three-eighths inch above the tops of the first blocks, and the lower end of the third pair of blocks are the same distance above the upper ends of the second pair. A frame so prepared will take three bars of cells and have about a two-inch space for comb beneath the top bar, although the comb is not necessary. A strip of tin tacked on one side of the frame end-bars will keep the cell bars in line as they are slipped into the frame.

To attach cell cups to the bars, some melted wax is ladled along the bar to form a base for the cups. Each wax cup is picked up by inserting the forming stick into it. The basal end of the cup is dipped into the melted wax, and the cup bottom is then pressed lightly into the wax on the bar. The cups are spaced about three-

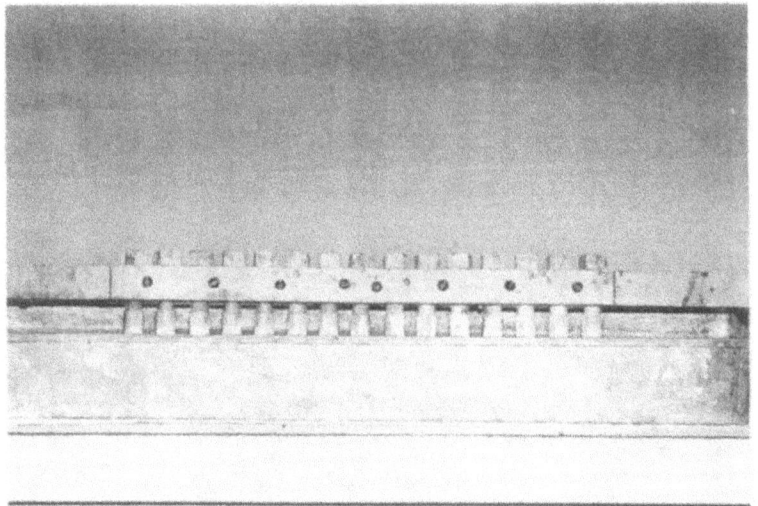

Figure 38. Dipping forming sticks in cold, lightly soapy or starchy water.

Chapter III

Figure 39. The forming sticks are dipped into melted wax to a depth of about 5/16 inch, and are immediately withdrawn. They are held momentarily above the dipping pan while the wax on the sticks partially solidifies before they are dipped again. From *Contemporary Queen Rearing*, Dadant and Sons, Inc.

quarters to seven-eighths inch from center to center; 15 or 16 cups are placed on each bar. When a bar is filled, melted wax is ladled along each side of and between the cell cups, reinforcing their attachment to the bar and making a firm base by which to handle the cells when the bees have completed them.

• *Quantity Production.* When cell cups are needed in large quantities, dipping one cup at a time is impractical. A form for dipping many cups at once can be made by attaching a number of cell forming sticks to a wooden block or a strip of wood, as was done by W. H. Pridgen (1900). A convenient form (Fig. 39) has a row of 15 or 16 forming sticks spaced three-quarter inch apart from center to center. The sticks are all the same length. Two of these racks of framing sticks are required for most efficient cup production. The forming sticks must be kept clean and polished. They can be cleaned by dipping in boiling water for a few minutes, and may be polished with a soft cloth.

Pure, clean beeswax is needed for making the cell cups. It is most conveniently melted in a water-jacketed tray, or a thermostatically controlled electric frying pan that is long enough to permit the entire row of forming sticks to be dipped. Adjustable guides at the ends, or blocks adjacent to the ends, set the depth to which the cups are dipped. The temperature of the wax is maintained at just above its melting point. The wax in the dipping tray is kept at proper level by frequent additions of wax chips. A narrow two-by-three inch deep tray long enough to receive the entire row of forming sticks (Fig. 38) is also required. This tray contains dilute, bland, soapy water or thin, corn-starchy water. A large container of cold water available for washing the cell cups as they are pushed from the forming sticks is useful.

To make the cups, the sticks are first dipped into the soapy water and are then shaken, or are touched to a towel. to remove excess water. They are dipped about five-sixteenths inch into the melted wax in the dipping tray and are immediately withdrawn and held above the tray until the wax on the sticks has partly solidified (Fig. 39). This dipping process is repeated four or five times.

After the last dip, wax is ladled onto a cell cup bar that is laid over a second but shorter tray of beeswax. With the dipping sticks steadied by a support (Fig. 40), the bases of the wax cups are rested in the wax on the bar, and melted wax is ladled along one side and between the bases of the wax cups.

With the forming sticks still steadied against the support, let the wax partly solidify and then lift the dipping sticks and the adhering bar from the ladling tray and lower into the large container of cold water. While the cups are held under water the forming sticks are retracted from them by evenly pushing the cell bar away from the forming sticks, as illustrated in Fig. 41. The cups are at-

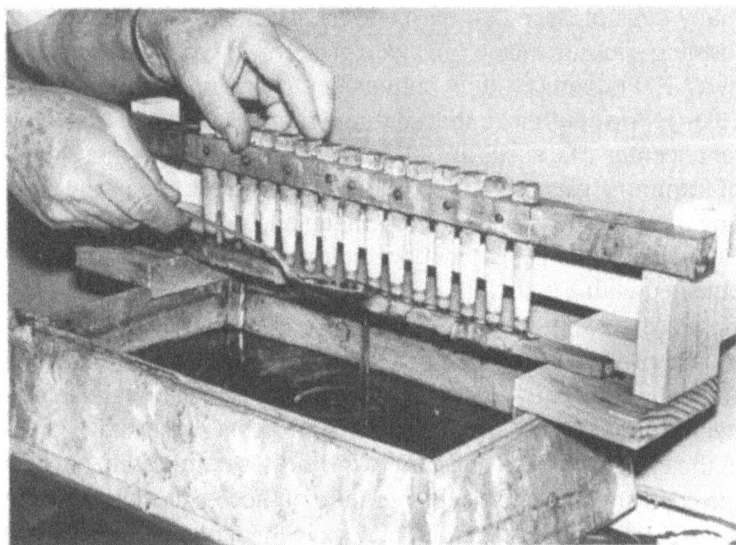

Figure 40. After the last dip, the cups on the forming sticks are rested on a cell bar, and with the rack of forming sticks steadied by a support, wax is ladled along and between the cell cups.

Chapter III

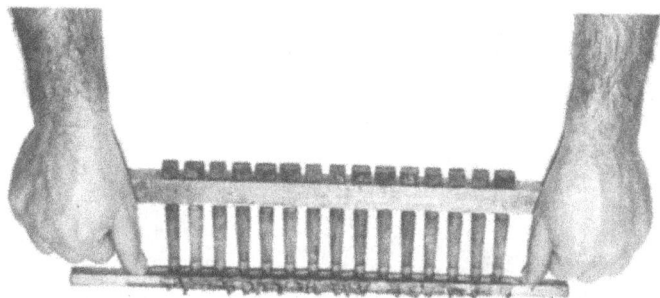

Figure 41. The cooled cell cups are pushed off the forming sticks by even pressure on the ends of the cell bar.

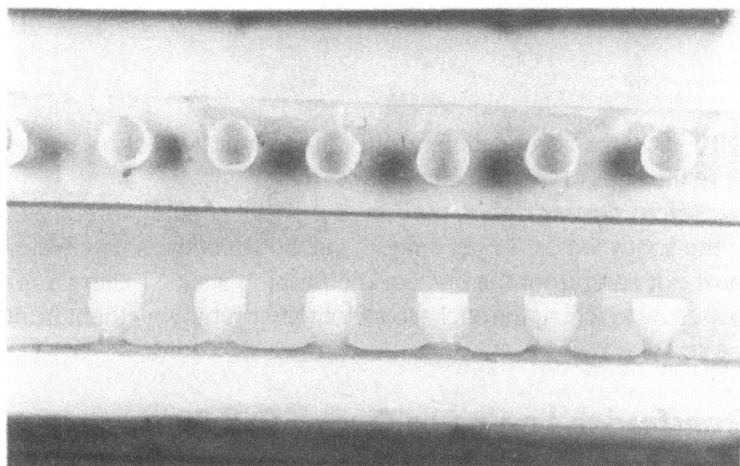

Figure 42. Dipped cells are attached evenly to the bars, and they have a thick wax base for handling individual cells.

tached to the bar and properly spaced; the cell bottoms are concave and smooth (Fig. 42).

The rate of cell cup production can be greatly increased by using two sets of forming sticks in rotation. To begin, one set of sticks is dipped. After the last dip, and guided by a support rack, the bases of the cell cups are rested on a cell bar laid over the tray of wax, and wax is ladled along the side of the cups and between them. The wax on the cell bar will solidify around the cell cups that are formed on this set while the other set of forming sticks is dipped. This second set, after dipping, is set aside until the cell cups of the first set of sticks are pushed off the forming stick. The

Figure 43. A grafting house in the apiary of cell builders is very convenient. From *Contemporary Queen Rearing*, Dadant and Sons, Inc.

cell cups of the second dipping stick are then attached to a bar as were the cups of the preceding bar.

Before dipping again, the forming sticks should be returned to the soapy water. To get smooth cell bottoms the soapy water must not hang from the ends of the forming sticks; it can be removed by touching the sticks to a cloth after removing them from the soapy water.

Transferring Larvae to Queen Cell Cups (Grafting)

Attention should be given to the temperature, light, and humidity of the room in which the grafting is done if a large number of cells is needed. The temperature of the room should be at least 65°F, and about a minimum 50% humidity is recommended to prevent dehydration of larvae, or of royal jelly if it is used to prime the cells. Humidification can be provided by a pan of water on a stove, by sprinkling the floor with water, or by hanging wet cloths in the room. Damp cloths laid over the grafted cells will aid in preventing the drying of the brood food and larvae even if the humidity of the room is low.

Good light is essential. It should be bright, and so located that the larvae can be seen clearly and be picked up and transferred quickly and without injury. Sunlight can be used, but wind should be avoided. A fluorescent lamp makes one of the most satisfactory lights for all weather conditions. Miners' electric lamps or "snake" lamps are very handy.

The proper conditions of warmth, humidity, ventilation, and light can best be provided in a small grafting house located near the cell builders (Fig. 43). Though it need not be large, enough space is needed for a small work bench and stool, storage for cell bars and the frames to hold them, and a wall area for a record-work chart showing the date of graft, the origin of the larvae, the cell builder colonies, and the date and the disposition of the cells. A dependable source of heat and light is standard.

If the weather is cold or windy, the combs of larvae and the grafted cells should be protected between the colonies and the grafting place by carrying them in a covered comb carrier or box, or by placing a towel over them. The experienced beekeeper who has bees in many locations, and raises cells within each apiary to requeen the colonies of that apiary may slight these precautions as the colonies are worked. The production of excellent queens demands organized and disciplined care, however, and major deviations from good practice may be reflected in poorer quality queens.

• *Grafting Needles and Grafting.* Two main types of transferring or grafting needles are in use, the *straight* and the *automatic* (Fig. 44). The former can be constructed from a six-inch piece of 14-gauge wire (about the size of baling wire). A jelly spoon with a large rounded end is handy in gathering jelly from queen cells. A needle with a small flattened end bent upward about 20° is used by some queen rearers.

The usual straight needle has one end thinned and expanded, and its tip rounded. One-quarter to one-half inch from the end it is bent to an angle of about 30°. This is the "jelly spoon." The other end, or another wire, is flattened and tapered to approximately 1/32 inch wide with the point made very thin. The terminal 1/16) inch is bent at an 80° angle to form the lifting hook. The underside and corners of this hook are rounded so that the hook does not dig into the cell bottom when it is slipped under a larva. The terminal 1/2 inch above the lifting hook is bent away from the lifting hook at a 30° angle to give unobstructed vision into the

Figure 44. Grafting needles, jelly spoon, coarse grafting needle, fine grafting needle, offset grafting needle, automatic needle, Chinese needle and 000 artist brush.

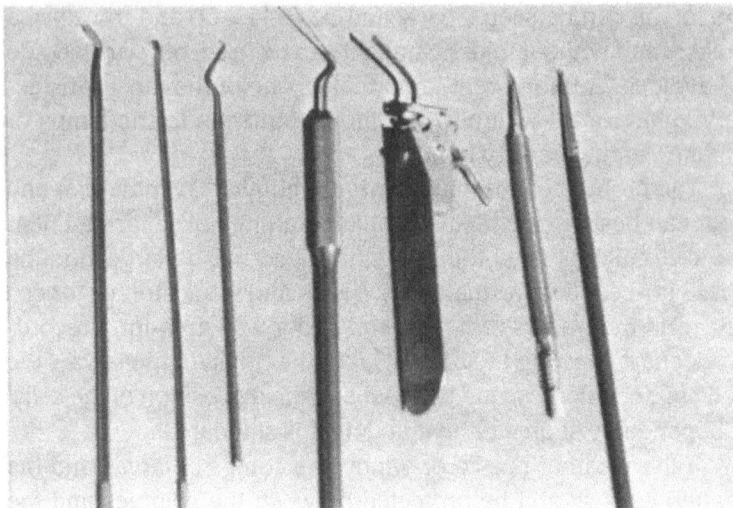

cell while the hook is being slipped under a larva. Talented bee-keepers have designed, and made, refined versions of the straight needle, one of which has the small, flat, terminal lifting hook deflected to the left side as viewed from above. Royal jelly spoons are handy for collection of jelly from 2-3 day old queen cells for use in grafting when jelly is used.

The automatic grafting needle has a watch-spring "tongue" that slips beneath the larva and its jelly in the worker cell. The tongue is retracted by a small coiled spring that leaves the transferred larva in the bottom of a cell cup on a minute bed of jelly. The tongue is operated by a lever mechanism attached to the end of a thin strip of metal that can be held between the fingers. A needle made in China, that functions in a similar way, is tubular with an extendible, flexible blade of hornlike material that slips under a larva and its jelly. Some beekeepers prefer to transfer the larvae with a 00 or 000 artist brush. Rarely, use is made of a grafting needle, handmade from a chicken wing-feather as was done more than a century ago.

In using a straight grafting needle, the cell cups are sometimes "primed" with royal jelly (Fig. 45). Priming the cups facilitates removal of the larva from the needle tip by floating it off. It also furnishes food for the larva, and retards the larva's drying. For the first graft, jelly can be obtained from two-day old dry grafted cells or from natural cells. The larva is removed from the cell and the jelly diluted with a drop of warm water and stirred until it has

Chapter III

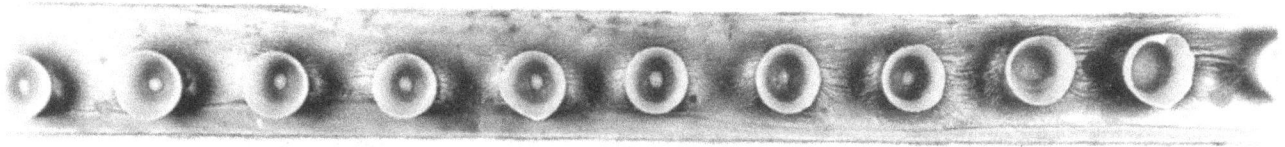

Figure 45. Primed cell cups.

an even consistency. A portion of the diluted jelly is taken up with a jelly spoon, from which an amount about the size of a large pin head is scraped off and deposited in the center of the bottom of the wax cup (Fig. 45). Another, and easier, way to put the jelly into the cups is with a medicine dropper.

Dry grafting is the usual method of grafting that is employed by commercial queen producers. The cell cups are not primed, and larvae are grafted "dry" with a straight needle. The cups are warmed to soften the wax so the needle can be pressed easily and slightly into the cell bottom and be withdrawn from under the larva. The automatic needle deposits the larva and its jelly on the cell bottom, and the cell cups do not need to be warmed.

Daylight is a usable source of light, but electric light is better. The operator should be seated, and the position of the comb adjusted until the larvae can be easily seen . With the bar of cell cups resting on the comb (Fig. 46) or held in the fingers, the point of the straight grafting needle is carefully inserted into the jelly beneath a larva by sliding the needle down the side of the cell wall and slipping the point sideways under the larva. The larva is lifted out of its cell and is transferred to the top of the jelly in the cell cup. The larva is actually floated off the needle and onto the jelly and is not immersed in the jelly.

When the automatic needle is used (Fig. 47), the cups are not primed because this needle picks up considerable jelly with the larva, and the jelly and the larva are deposited together in the

Figure 46. Grafting. Transferring larvae from comb cells to queen cell cups. A straight needle is being used.

Figure 47. Grafting. The automatic needle is being used.

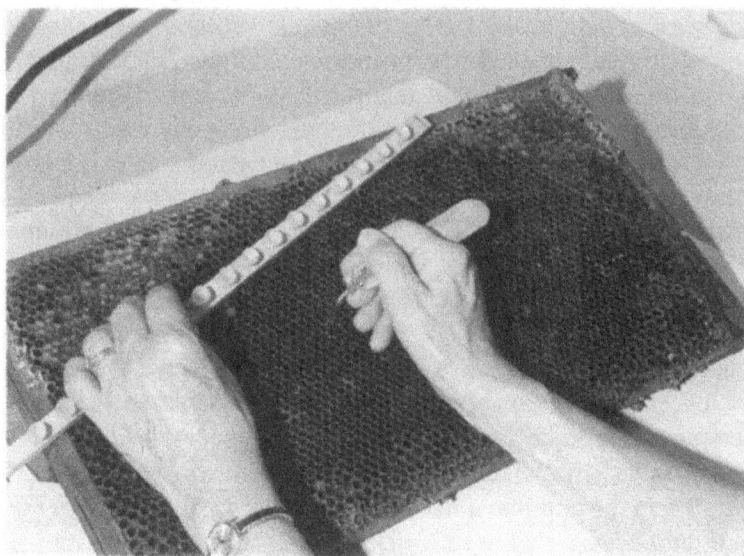

bottom of the cell cup. The tubular extension is lowered along a side of a cell to the bottom, and the lever is depressed so the spring tongue protruding from the end of the extension can be slipped under the larva and its bed of jelly. With the lever still depressed, the larva and the jelly are lifted from the cell and lowered to the bottom of a cell cup. Release of the pressure on the lever allows the spring tongue to withdraw, depositing the larva on its jelly on the cup bottom.

If the process of transferring the larvae is slow, each bar should be put into the cell builder as soon as grafted. If the process is rapid, three or more bars can be grafted before they need to be put in the cell builder. Care should be taken, as previously noted, to keep the larvae from becoming chilled or dry during the transferring process. The sooner the larvae receive the care of the nurse bees the less likely they are to be injured.

• **Double Grafting.** Double grafting is the removal of the larvae from accepted cells one or two days after grafting and replacing them with larvae 24 hours old. This, conjecturally at least, provides the larvae with abundant food from the time of transfer and generally ensures that considerably more food is put into the queen cells than the larvae can eat. Care should be exercised in the second transfer when cells are double grafted. Unless the larvae are removed without materially changing the consistency of the royal jelly, the bees may remove all of the jelly and start over again, thus eliminating any beneficial effect the double graft might possibly have.

The Cell Builders

Queenright Cell Builders

Cell builder colonies, in queen rearing operations, are intermediate in function between the selection of breeder queens and the mating of the virgin queen. It is in their care that the caste-undetermined female larva that hatched from the egg becomes a queen and emerges from her queen cell as an adult queen. This care is a colony effort, with bees of appropriate age and development cooperating.

Cell building is a seasonal activity as a part of colony reproduction, i.e., swarming. Cell building is also a colony maintenance response (supersedure) to the aging of the colony queen. A third incitement to rearing a queen is loss of the colony queen. The queen rearer, by inducing the colonies to rear queens from larvae given them, exploits these behavioral characteristics of colonies, and endeavors to provide and maintain a suitable cell building environment.

Cell builders are either queenless or queenright, and may be starters, finishers, or both starter and finisher. Each type is composed of abundant bees of all ages and is especially strong in nurse bees. Each cell builder is well provisioned with honey and pollen and is fed sugar syrup continuously for stimulation. The colony is arranged so the nurse bees are concentrated in the cell-building area.

Larvae prepared by the Miller or Alley methods will be accepted and completed most successfully by queenless colonies or in a swarm box. Grafted cells will be accepted by either queenless or queenright cell builders, or in a swarm box.

• *Queenright Finisher or Starter-Finisher.* The two-story queenright cell builder is the basic "work-colony" of the queen rearing industry. Queenright cell builders are used extensively for finishing cells that are started in swarm boxes or starter hives, and they are favored by hobby and sideline beekeepers to start and finish cells. The queen is confined in the lower of the two broodnest bodies by an excluder and the cells are built in the second story.

Queenright cell builders, like all cell builders, must be strong. This makes swarming somewhat a problem, but is one that can be minimized by maintaining a shallow body of dark combs below the active broodnest on the bottom board as clustering space for foragers who have little, if any, in-hive activities (Fig. 27). Honey bees do not normally store pollen or honey below the broodnest but store honey and pollen above and beside the brood area with the pollen next to the brood and the honey outside the pollen. Young queens are less inclined to swarm than are older queens and they should be used in cell builder colonies. A frame of foundation placed off-center in the lower broodnest body with the queen damps swarming by supplying work for bees secreting wax and building new comb for oviposition. Observation and experience reveal that bees with ample suitable space to engage in the activities appropriate to their age and development are less likely to swarm than are bees with severely limited space for their activities. A clipped queen will prevent loss of the bees if the colony

Figure 48. Three- and two-story queenright cell builder colonies that can function as starter-finisher colonies or as finisher colonies that complete the cells started in the starter colonies or in swarm boxes.

should swarm; the bees return to the hive but the queen is lost on the ground.

• *Two-Story Queenright Cell Builder.* The two-story cell builder is suitable for hobby, sideline, or commercial queen rearing operations, and is recommended when a few cells are to be produced by the grafting method. Strong colonies with two brood nest bodies are selected for cell builders. Emerging brood may be added to bring them to the needed strength. The colonies should be fed for two or three days prior to use as cell builders, and it is recommended that the cell builders be made up at least one day before giving them the first lot of cells.

In making up a two-story cell builder, the queen is left below an excluder in the lower body with emerging and sealed brood and an empty comb. Unsealed brood is put into the second body above the excluder to bring the nurse bees to the cell-building area. A comb of honey and pollen is placed next to the two side walls of the upper body and the unsealed brood is put toward the center with the youngest in the middle, where the first cells will be put. A frame containing pollen is also placed near the middle so it will be readily available to the nurses. Remaining spaces, if any, are filled with emerging brood from other colonies. The cell-building colony is fed *continuously* with 50/50 sugar syrup (50 parts sugar/50 parts water by volume).

Approximately 15 minutes to one hour prior to giving the first graft of 3 or 4 bars of cells (45-64 grafted cell cups) to the cell builder, the comb of larvae at the center is removed, leaving a comb of young larvae on one side of the space and the pollen comb on the other. The nurse bees, which had been feeding the larvae that were removed, cluster in this space and are eager to feed the cells given to them. Thereafter, new cells are given every third or fourth day in the same place as the preceding graft, which was moved to the opposite side of the comb of larvae.

The arrangement of placing newly grafted cells between a frame with young larvae and a frame with pollen should be maintained. By the time each new graft is given to the cell builder the cells of the preceding graft will be at the sealing stage and no longer requiring the attention of the nurses. They are moved toward the side; after the cells are sealed they are merely incubated by the cell building colony, and their location in the hive is not important. It is advantageous, however, to establish a definite routine so that cells of different ages are located in specific relation to each other. Such a system eliminates searching for the oldest cells and guards against leaving ripe cells to emerge in the cell builder.

When a three-day interval of grafting is used, the nine-day-old queen cells may be removed from the cell builder at the time the third graft is given. The frames of cells can be placed in an incubator and prepared for the nuclei the following morning without again disturbing the cell builder. The incubator temperature should be kept at about 93°F and the humidity around 50%. An open pan of water in the incubator will keep the humidity at a satisfactory level. Care in maintaining a suitable temperature for the developing queens is critical at this time. Any sudden or prolonged chilling will tend to retard the growth of the queens and could result in underdeveloped wings.

Some queen breeders prefer to give their cell-building colonies two or three bars of cells every fourth instead of every third day, as just described. In that program, the last grafted cells are generally ready to be sealed and are simply pushed toward the opposite side of the hive, with a frame of young brood placed between the last frame of cells and the one with newly grafted cells. Ten days after the first graft is given to the cell builder, the first frame of cells must be removed and distributed to nuclei or other colonies, or be caged in a nursery colony. At this time the condition of the colony below the queen excluder is checked for queen cells, and brood is raised, bees added, and the colony otherwise brought up to its peak of queen-rearing efficiency.

• **Three-Story Queenright Cell Builder.** The three-story cell builder is merely a two story cell builder with a body of brood and honey inserted between the two bodies above the excluder (Fig. 48). The cells are produced in the top body. In some cases, the queen cells are placed in the second story of a three-story queenright (or queenless) cell builder on the assumption that the super of honey and bees above tends to stabilize the temperatures during spring or early summer.

• **Queenright Finishing Colonies** are actually two-story queenright cell builder colonies that are used for the one purpose of caring for the growth and development of queen cells that were started elsewhere, such as those started in swarm boxes.

Maintenance of Queenright Cell Builders

The queens in queenright cell builders usually maintain a good egg-laying rate and thus keep the colony strong. If more bees or brood are needed, they are supplied from other colonies. The brood that emerges below the excluder gives room for the queen to lay, and the young bees are attracted to the body above by the larvae that were moved there. About four days after the colony is rearranged new larvae are hatching in quantity below the excluder and more nurse bees are required in the lower body. Most of the larvae which were put above are sealed by this time, so fewer nurse bees are required to care for the young larvae which were raised above the excluder. Thus, there is a reversal of nurse bee concentration. It then becomes necessary to raise a comb of young larvae from below, exchanging it for a comb of emerging brood, or of honey. The young brood is put next to the youngest queen cells in the center of the second body. This exchange in the position of the larvae and emerging or sealed brood is made before every new graft.

The Queenless Cell Builder

Queenless cell builders are one- or two-story and are used to start, or to start and complete cells. The one-story queenless cell builder is recommended when queen cells are produced by the Miller method. These colonies cannot maintain themselves, and they need supporting colonies for brood and worker supply. Support colonies have a two-body broodnest with the bodies separated by an excluder. Sealed brood is raised from the lower broodnest body into the upper one for convenience in obtaining bees for the queenless cell builders without searching for the queen of the support colony.

Before queenless cell builders are established, their parent colonies are fed light sugar syrup (50/50) for at least a day or two. Pollen is given in bee-packed pollen combs. Pollen can also be fed in cakes, in candy, and by filling the cells of a comb half full of pollen pellets and spraying the pellets with light sugar syrup. Light sugar syrup is used because it is more like nectar than is heavier syrup.

In making up a one-story queenless cell builder the selected colony is dequeened, leaving sealed and emerging brood except for two frames of young larvae which are placed in the middle of the hive. A pollen comb is placed beside one of the combs of larvae. The combs next to the walls of the body should be well filled with pollen and honey. If necessary, more bees are added to make the colony strong. The colony is fed continuously unless there is an ample and reliable nectar flow in progress.

The first cells are given to the colony the day after it is made queenless. Shortly before the first cells are given, one of the center combs of larvae is removed, leaving a space with a comb of pollen on one side and a comb with larvae on the other. The nurse bees cluster in the space. Succeeding grafts, or prepared cells, are put in the center at three- or four-day intervals, the older ones being moved to the side as with queenright cell builders. Alternatively, the day old cells are transferred to a queenright finishing colony.

Queenless cell builders are maintained by regularly adding bees that were shaken from other colonies into cages, or by replacing combs in the cell builder hive, from which bees have emerged, with combs of sealed or emerging brood from the supporting colonies. This is done once or twice a week. The comb of larvae next to the grafted cells should be replaced with younger larvae a few hours prior to each graft, and the sealed queen cells shifted to their new location in the hive. The nearly mature cells are removed the ninth or tenth day after grafting and are distributed to nuclei or caged in a nursery colony.

If it is necessary to add more bees, they are added before a new graft. The additional bees, from an outyard if possible, are shaken into a screened bulk bee cage (Fig. 66) that is supplied with a comb of honey, or they are shaken into shipping or holding cages from brood nests of supporting or other colonies an hour or more earlier, and fed. Several hours later they are sprayed lightly with water or dilute sugar syrup (1 part sugar/2 parts water) just before they are added to the cell builder by letting them run into the entrance. Bees can be added to queenless cell builders by allowing them to run from the cages into the cell builder hive entrance in late afternoon. A ramp the width of the entrance temporarily placed from the entrance to the ground or bench is helpful to guide the bees to the entrance.

• *Two-Story Queenless Cell Builder.* The two-story queenless cell builder is well supplied with honey and pollen in the first story, and the second body is made up and maintained like the one-story cell builder. Nearly twice as many bees are required to fill the hive as is required for the one-story cell builder. The colonies are examined once a week and the natural queen cells are destroyed. Enough bees should be shaken from each comb to expose spurious cells which may be hidden in the comb corners. Overheating should be avoided in all cell-building colonies. When colonies are placed in the sun without shade boards, insulated tops, ramadas, or adequate ventilation, the temperature of the upper part of the hive may rise so high at times that many of the bees may leave, crowding into the lower hive region or hanging onto the outside. The cells will be neglected to a greater or less extent and poor queens often result despite a sufficiency of bees and otherwise excellent conditions.

During periods of heavy nectar flow, the bees often build comb on the sealed queen cells, particularly if the cells are made by the Alley method. This can be prevented to a large extent by putting a frame or two of foundation in the cell builder or by using narrow frames for the cell bars. The narrow frames (as wide as a frame bottom bar) permit the bees to cluster on the cells when the adjacent frames are close to it, but do not provide space enough for the bees to build comb on the sides of the cells.

Cell-Starter Colonies

Queen cells can be started and finished in the same colony. This method is used commonly by commercial queen breeders although many prefer to start the cells in one colony and finish them in others. Either of two types of starting colonies are suitable— the queenless starting colony, or the swarm box.

• *Queenless Starting Colony.* This starter colony is a regular queenless cell builder used to start cells that will be built in other colonies. About an hour before the cells are to be given to the colony, the combs of larvae are removed. When the cells are grafted, they are put in the places that were occupied by the frames of larvae. Two frames with grafted queen cell cups (90 to 128 cells) may be given at one time. They are removed 24 hours later and distributed to the finishing colonies, usually one to three bars of cells to each finisher.

This type starter colony will start two or three successive grafts of cells, but as the bees grow older their efficiency as nurses decreases and fewer cells will be accepted and fed adequately. Thus, starter colonies must have new bees and/or brood added every two to four days.

Swarm Box type of cell starter (Figs. 32, 33) usually holds five standard frames, with a three-to-six-inch space below them that is screened on the sides. There is no entrance, and the bees are confined the entire time they are in the box. A division board feeder with sugar syrup is placed next to each sidewall of the swarm box and a comb of pollen is put in the center space. The box is now stocked with five to six pounds of nurse bees from the upper brood nests of donor colonies that have young larvae that were raised from below, and which were fed light sugar syrup for two or three days prior to their use as bee donors, unless there is a good nectar flow. After stocking, the box is set in a cool, dark place. Two hours later the bees are ready for cells. Some beekeepers fill combs with sugar syrup rather than use a feeder.

Two frames with three or four bars each (about 90 to 128 cells) are given at one time. The box is jarred to knock the bees down and the frames with the cells are put into the two empty spaces. Twenty-four hours later the cells are removed and distributed to the finisher colonies. When the cells are removed, the swarm box is *not* jarred and the bees are brushed lightly from the cells. Queen cells should *never* be shaken; queens with wing buds will often suffer damage so they are unable to fly. Two or three succeeding grafts of cells can be started by the same lot of bees.

• **Modified Swarm Box.** A cell starter in which the bees are confined above a two-story colony for twenty-four to thirty-six hours while cells are being started is an efficient and convenient modification of the swarm box.

A strong two-story colony is selected, provisioned with pollen, and fed sugar syrup continuously . The queen is kept in the bottom body below an excluder with sealed and emerging brood, pollen, honey, and empty combs. Combs of young larvae are moved to the upper body. This arrangement is maintained by moving young larvae to the top body, and recently-sealed brood to the body below. This should be done at three-day intervals if successive grafts are to be started.

A full-depth empty body is prepared with two well-filled combs of pollen in the middle of the body. A space is left between them for the cells, and a division board feeder is placed next to the outer side of each pollen comb (Fig. 49).

One or two hours before newly grafted cells are to be given to the starter, the prepared body is set upon a screen made of 8-mesh hardware cloth fastened between two 1/4" thick rims (Fig. 50). They are placed together on an upturned hive cover, or on a "super horse" (see Remarks) beside the donor colony. The top body of the donor colony, which has young brood, nurses, and wax builders, is set on top of the prepared body and the bees are shaken from the combs into the starter body, after which the brood body

Figure 49. Modified swarm box. The cell starter body has two pollen combs at the middle of the body with a space between them for the frame of grafted cells. A division board feeder is next to each pollen comb. The body is set on an 8-mesh hardware cloth screen that is framed with wood strips. There is no entrance to the body. From *Contemporary Queen Rearing*, Dadant and Sons, Inc.

Figure 50. Eight-mesh hardware cloth is placed between the top body of the modified swarm box and the prepared body and bees. From *Contemporary Queen Rearing*, Dadant and Sons, Inc.

is returned to the donor colony above the excluder. The stocked cell starting body and the screen beneath it are now placed on the top body of the donor colony (Fig. 51), and the feeders are filled with sugar syrup (1 part sugar to 1 part water by volume). There is no entrance to the starter body. A graft of four bars of cells is put into the space between the two pollen combs. Twenty-four to thirty-six hours later the started cells are transferred to a finisher colony. The screen is removed; the bees of the starter body are shaken down into the donor colony below; and the starter body is removed, or left on the hive. This an excellent cell starter and is convenient and economical to operate.

Cell-Finisher Colonies

Finisher colonies are regular two-story queenright cell builders used to finish cells started in starter colonies or swarm boxes. They are used in many, if not most, commercial queen rearing apiaries, and, when maintained, are one of the best cell builders for the hobby and sideline beekeeper for both starting and finishing . One to three bars of started cells (one day from grafting) each three days.

Figure 51. Modified swarm box colonies. From *Contemporary Queen Rearing*, Dadant and Sons, Inc.

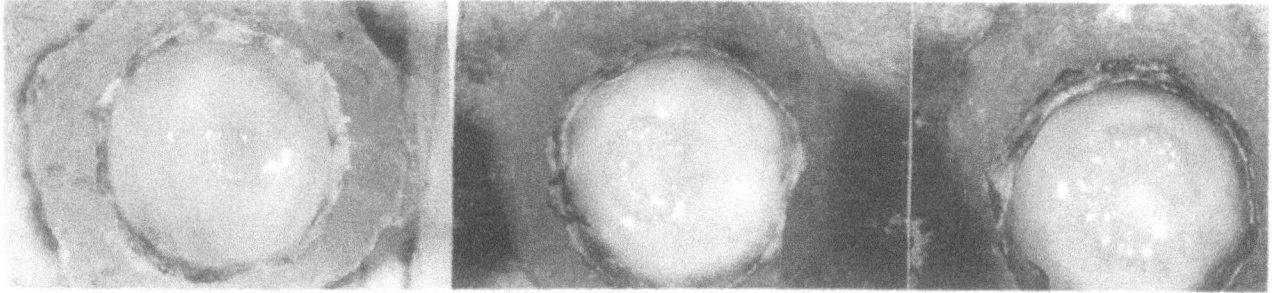

Growth of the Queen Larva and Queen Cell

The larvae are fed lavishly by the nurse bees for three days following grafting (Fig. 52) and they grow rapidly as the cell lengthens (Figs. 53, 54, 55). The feeding by the nurse bees is completed on the fourth day and the cells are sealed (Fig. 56). The larva continues to ravenously eat the abundant supply of royal jelly until the sixth day after grafting when it secretes a cocoon and turns its head down to the lower end of the cell and becomes motionless.

Figure 52. Growth of larvae in queen cells. Larvae one, two and three days following grafting.

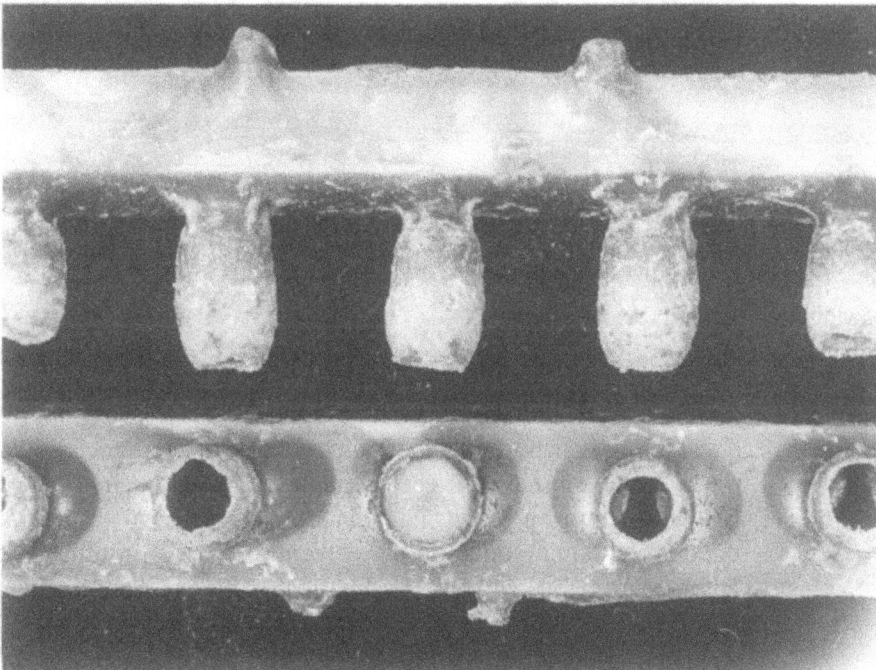

Figure 53. Queen cells one day following grafting.

Figure 54. Queen cells two days following grafting.

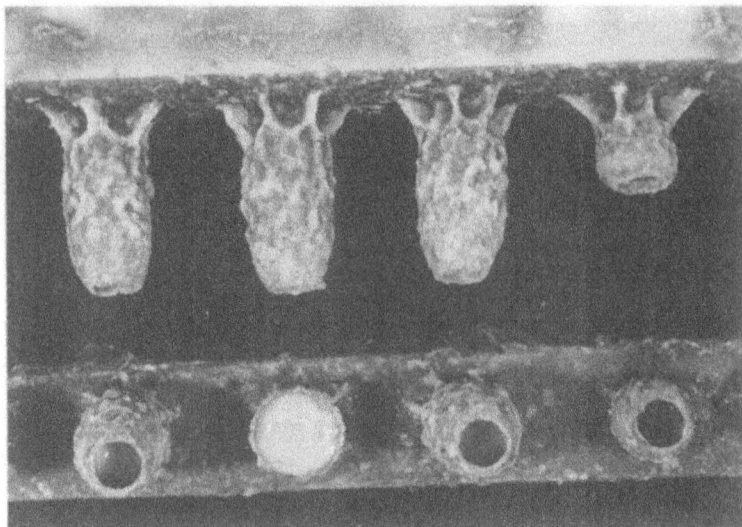

Figure 55. Queen cells three days following grafting.

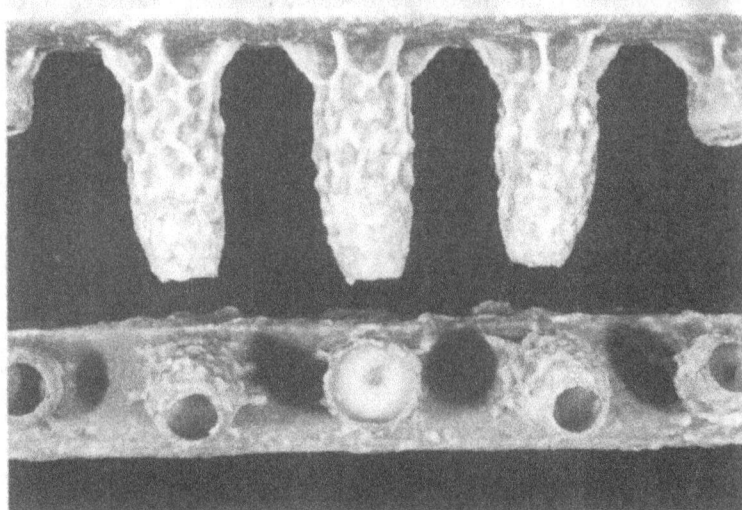

It is now a *prepupa* (Fig. 11). The prepupa remains such for two days during which time the larval structures are remodeled into those of the adult queen. The prepupa then *moults* (sheds it old body wall) to reveal a *pupa* (Fig. 12) that has the form of the adult queen. The pupa moults on the eleventh day after grafting and emerges from the cell on about the twelfth day as an adult queen (Fig. 14).

Chapter III

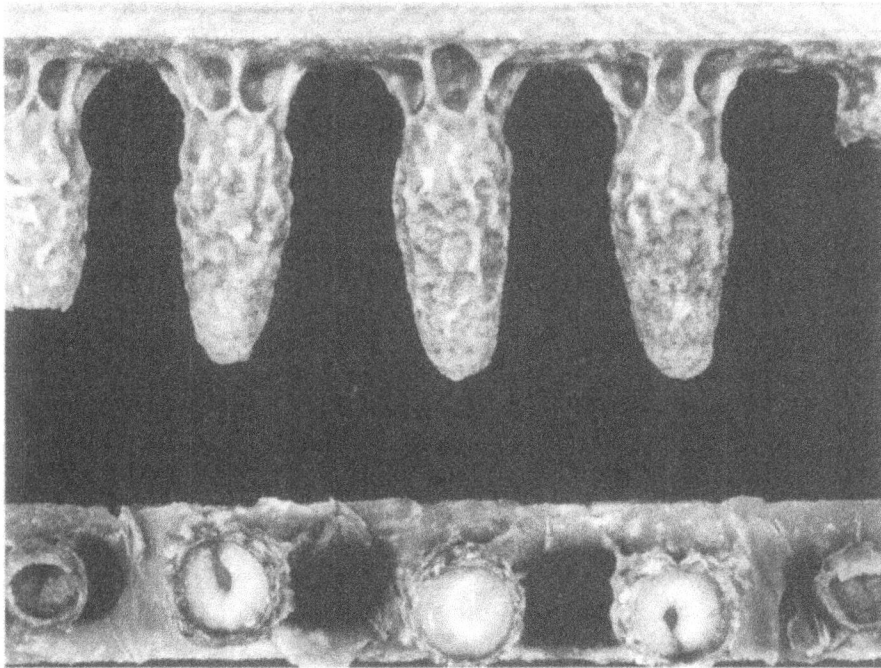

Figure 56. Cells sealed on the fourth day following grafting.

Records

Simple records in a notebook, on the hives, frames, or bars are sufficient when cells are raised in small numbers. A more complete automatic system is needed when large numbers of cells are produced each day. The method of putting new cells in a definite place in the cell builder and moving the older ones to the side shows the relative ages of the different cells within the hive. By dividing the cell builders into three groups, into which the grafted or started cells are put into a different group on successive days, a system of rotation is set up whereby each group receives new cells every three days and the "ripe", (nearly mature cells), are removed. All colonies of a particular group receive newly grafted cells the same day, and "ripe" cells are taken from them on the ninth day if they are placed in an incubator over night, or on the tenth day if they are distributed to nuclei. Since any particular group will receive attention on different days of the week during successive weeks, the record of each graft will show into which group of builders the cells were put and when they will come out.

These records should also indicate the breeding queen used and into which nuclei or group of nuclei the cells were placed if the queen breeder is to select stock for breeding purposes. The following type of record may be kept on a blackboard in the grafting house or on a sheet of paper—

Graft Record

Date grafted	No. of cells	Breeder	Cell builder	Cells accept	Date out	Remarks Disposition

The *Date out* is entered at the time the graft is made. When cells are taken out of the builder a line is drawn through the entry.

As many as three bars of cells (about 45 to 48 cells) can be put into a cell-building colony at one time, either for finishing or for both starting and finishing, and, if the colony is in proper condition, new cells can be given the cell builder every three days. Usually at least 75 per cent of the cells are accepted and satisfactorily completed. Thus each cell-building colony can be counted on to produce about 35 to 40 cells each three days, or at a rate of about 12 cells per day. The cells are sealed on the fourth day and need only incubation. A new graft can be given to the cell builder three days after the preceding graft because the feeding of the preceding graft is nearly finished.

A Shelter for Cell Builders

Some commercial queen breeders erect substantial shelters in which to place their cell-building colonies as well as to provide a suitable place for preparing the larvae under all weather conditions. Such conveniences are provided in separate buildings, in rooms, in additions to warehouses where needed equipment is stored, or even under ramadas or canopies. A room ten feet wide with two outside walls will provide adequate space for two rows of cell-building colonies and an ample aisle in between. For greater convenience in manipulation, the hives are placed on benches. Windows are arranged for exit from the building of bees that escape from the hives when the colonies are being manipulated. A work room for grafting or preparing the cells and larvae can be set up at one end. Heating units and a fan to distribute the heat evenly throughout the room may be desirable. A cell builder colony shelter, whether enclosed or not, makes it possible to graft at any time during the night or day, regardless of the weather. The prepared beekeeper is not troubled by wind, or rain, or darkness.

The Care and Handling of Queen Cells

Queen cells must be handled with care at all times. Development of the queen can be disrupted by heat or cold. A sudden jar on unsealed queen cells may dislodge the larvae from their beds of royal jelly; and the wings and legs of pupae in older cells may be injured. A comb containing queen cells should never be shaken if the cells are to be used. When cells are removed from cell builders the bees covering them are smoked and brushed off carefully.

If a ripe queen cell is held before a strong light and is tilted slightly to one side, the queen in the cell will rock, and the outline of her body within can be seen. Sometimes the movement of legs or wings can be discerned. Queen breeders "candle" their cells in this way to be certain of distributing to colonies or nuclei only cells containing living queens.

Young queens usually emerge from their cells twelve to thirteen days after hatching from the egg. This is true whether the queen develops from an egg laid in a preconstructed cell, in a cell built by the bees around a worker larva, or from a larva grafted into a queen cell cup and finished in a cell-building colony.

To remove the cells from a bar, the bars of cells are laid on a flat surface and the cells are separated from the bars by sliding a warm, thin, knife blade between the bars and the cell wax base. The cells are cut apart and are placed either in cell blocks with rows of holes for the cells (Fig. 57) or between layers of cotton or cloth in a box. The cells should not be exposed to a hot sun or to cool winds. Chilling will either kill the queen or delay her emergence, and sometimes results in wing deformities. Overheating is especially dangerous to unemerged queens.

The appearance of a queen cell indicates its approximate age. Mature queen cells are well sculptured and the bees usually reduce the thickness of the cap closing the cell. Shortly before her emergence from her cell, a queen can be examined, and even marked or numbered. By making a partial incision around the lower part of a ten-day old cell, the body of the cell can be tipped away taking the developed queen with it, and the queen can be dropped out into the palm of the hand. She can be reinserted into the queen cell, and the cell body returned to its normal position and stuck firmly to the lower part of the queen cell. The bees repair the cut and the queen will emerge normally.

The general practice in queen rearing is to remove queen cells from a cell builder and put them into nuclei one or two days before the queens are due to emerge. In this way, the cell serves as an introducing cage, permitting the nucleus to become accustomed to the cell before the queen emerges. If the cells are left in the cell builder longer, a queen may emerge before the cells have been removed. One of the virgin's first acts after emergence will be to tear holes in other queen cells and to sting the inmates to death (Fig. 16). She will pay little or no attention to unsealed queen cells or to those recently sealed; the workers will discontinue their construction and destroy the other sealed immature cells.

Nursery Cages

Sometimes cells are held in nursery cages in nursery colonies until the cells are put into nuclei or until the virgins emerge. These cages (Fig. 58) may be made of wood and wire or of wire cloth. The Alley wood-and-wire cage is probably the most widely used and it is economical to make. Two-hole and three-hole Benton queen shipping cages can be converted to nursery cages by enlarging the entry hole of the cages to accept a mature queen cell. Plastic cylindrical hair curlers can be used as nursery cages, and they are popular in Europe. Other cylindrical or rectangular cages have been described that are equally suitable. The cages need not be provisioned with candy; the bees of the colony will feed the queens. A small ball of queen-cage candy in the cage may be reassuring, however. Nurse bees are not needed inside the cages to care for either cells or queens.

Queens kept for even short periods in nursery cages that have a side of wide mesh screen wire may suffer damage to their feet. The large, central lobe of the foot, the *arolium*, may be partially, or completely, destroyed by worker bees that are outside the cage but are biting the wire. The bees can grab the queen's feet with their mandibles through the mesh spaces of large mesh wire screen. Some moderate damage to the feet may or may not adversely affect the queen's performance and longevity. Queens kept in nursery cages for extended periods, however, may have some

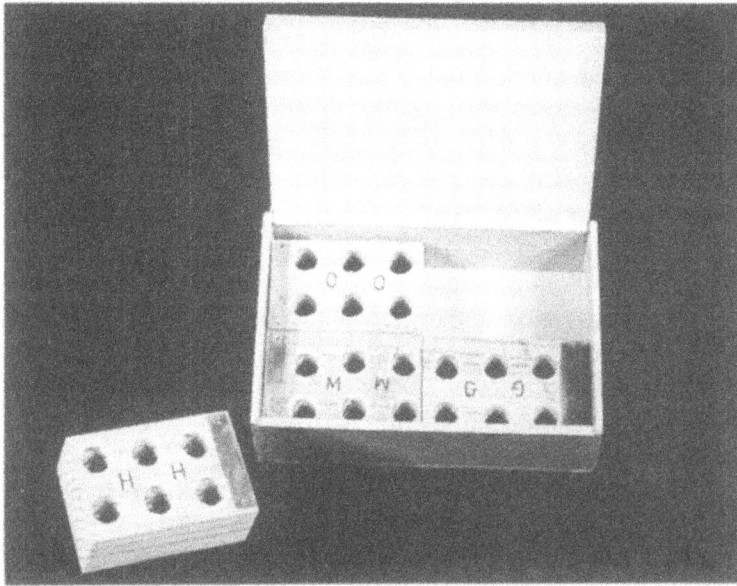

Figure 57. Cell block for transporting cells to nuclei.

or all of the tarsi chewed off making it impossible for the queen to grasp vertical combs. Damage can be prevented by using screen with finer mesh than that of window screen. A double layer of screen should be used when the cage is made with large mesh wire cloth. These practices have been in use for many years. Most queen breeders, however, prefer to leave the cells with strong cell-finishing colonies until they are ready to be placed in mating nuclei. If the mature cells are to be transported in cool weather, some method of keeping them warm should be used. Some bee-keepers use bees for this purpose. About a pint of bees is put in a honey pail that is ventilated by a few small holes. The cells are placed in the pail on layers of cloth or quilt. Other beekeepers place a warmed and insulated piece of iron or stone beneath the cloth. Too much heat, however, is as bad as, or worse than, too little.

When the identity of each cell must be known, the cells can be cut from the cell bar and placed in holes in blocks that are numbered or lettered to take to the nuclei (Fig. 57).

Figure 58. Nursery cages. Match-box cage (left), two-hole shipping cage with entry hole enlarged (center), and Alley nursery cage (right).

The Production of Queen Cells

Placing the Queen Cells in Nuclei

Queen cells are often hung between the top bars of two frames in the immediate vicinity of brood, or are pushed gently into the surface of the comb near brood or where the bees are clustering. If the weather is cold, the cells should be placed about an inch below the top bar of the frame so the bees can cluster around the cells and keep them warm.

In all cases, care should be taken to leave the end of the cell free for the queen to cut her way out. The cells should not be even slightly crushed at any time. If the cell has to be pushed into the comb it is advisable to first crush a spot on the comb slightly with the fingers for the location of the cell. The bees will fasten the cell in place.

A base of wax with which to handle the cell and to hang it between the frames or to serve as an anchor in the comb is preferable to either wooden cell cups or wooden chips. The bees will cut down the wax base after the queen has emerged, whereas the beekeeper must remove the wooden cell cups or chips.

Commercial Production of Royal Jelly

Smith (1959) described a method of producing royal jelly in quantity for research purposes. A small number of beekeepers throughout the world raise royal jelly to supply a limited demand for it for therapeutic uses. The methods are basically the same as those incorporated in the production of good queen cells for rearing queens.

A strong queenright colony can care for repetitive grafts of 45 cells. At the end of the third day, the grafted cells are removed from the cell-building colonies, cut down to approximately the level of the royal jelly by means of a sharp, hot, thin-bladed knife, and the larvae are removed, without injury, with fine forceps. The royal jelly is then removed by suction, with the vacuum arranged so the royal jelly is deposited in a bottle or collection tube of glass. As used by Smith, the tube is three-quarters to one inch in diameter, six to eight inches long, and open at both ends. One end is closed with a solid cork that fits snugly into the tube. In the other end, a cork is fitted with two glass tubes, one connected with the vacuum pump and the other being used to remove the royal jelly from the cells. The tube can be completely emptied by removing the latter cork and forcing the solid one through the length of the tube.

Bits of wax and moulted skins of the larvae are strained from the royal jelly by forcing the jelly through a 100-mesh nylon cloth held tightly across the end of the tube. The jelly should be refrigerated immediately.

The cell cups are reusable. Plastic cups work as well as those made of beeswax and may be preferred by some beekeepers.

MATING THE VIRGIN QUEENS

Mating the virgin queens is the last step in producing young laying queens. The newly emerged queen, though she is a fully developed reproductive female, has yet to attain the reproductive capability of a colony queen. She must mate and lay eggs before she can be termed a "laying queen". The usual period between emergence and mating is five to eight days, and the interval between emergence and oviposition ranges from nine to fifteen days. Some queens may lay earlier and some later than the normal range. The queen's instinctive impulse to mate is strong when she is six or seven days old and she will mate from any cluster of bees that is organized as a colony.

In the natural succession of queens in a colony the colonial environment is suited to progressive development of the queen's reproductive power. Queen rearing environments arranged by beekeepers are sometimes more variable, and, though virgin queens 24 hours old or older are remarkably hardy, the consistent production of the very best queens requires adherence to meeting the needs of the developing queens. The first consideration is to have enough bees of different ages to maintain a suitable environment for the cell and the queen under prevailing weather conditions. The second consideration is to have sufficient food present in the nucleus, or supply the food, to enable the nurse bees to provide the queen with proper nourishment for her reproductive development, and later for oviposition. The third consideration is to ensure an abundance of mature and suitable mates for the virgins. The greatest yield of vigorous, laying queens is derived from well-cared-for cell building and mating colonies, and that should be a goal of all queen breeders.

Mates for the Virgin Queens

One of the most neglected elements of queen rearing is the provision of suitable mates for the virgins. Each virgin mates on a mating flight with seven or more drones while flying; and more than one mating flight may be taken. Research studies have shown that there are "mating areas" within flying distance of apiaries that are recognized by both queens and drones, and that numerous matings take place there.

It is of paramount importance to the queen producer, and to the beekeepers who buy the queens, that *an abundance of drones of a designated stock be present at the mating area and surrounding areas to ensure complete inseminations and with drones of the desired stock.* One full drone comb has about 4150 total cells on the two sides, and may reasonably be expected to produce about 3500 mature drones. Allowing 20 drones for each virgin queen, the drones from one comb of drone brood would supply the mates for 175 virgins.

Sixteen days after the first drone comb is given to a drone-mother colony a second comb should be added to furnish the mates for the virgins following the first group that have laid eggs and were removed from the nuclei. A second drone colony added to the drone-mother colony group is in some cases preferable to adding extra combs, especially when feral, or other beekeeper -managed bees, are abundant.

The placement of the drone mother colonies within a mating apiary is not important. The queens and drones fly about the same distance, and are attracted to the same mating area. The "conscientious" queen producer will not rely on feral or neighboring drones to inseminate his virgins but will rear drones of the proper stock in numbers sufficient to saturate the mating areas. A drone comb in the broodnest of every colony headed by queens of selected stock, and two drone combs in the brood nests of breeder and potential breeder colonies is *not* exorbitant. The day after a spell of non-mating weather during which virgins emerged and are six days old or older, will be marked by the complete disappearance of mature drones in a mating apiary unless drones were present in enormous numbers.

Queen Mating Colonies

Because of queens' mutual animosity and workers' intolerance of multiple queens in a colony, queen rearers must provide separate colonies for the queens' post emergent development. The design, size, and construction of the mating hives vary, but these attributes appear to have no observable effect on the welfare of the queens. They do, however, reflect preferences of the queen rearer who is influenced by the fact that the mating and shipping of queens are the most costly operations of queen rearing because of the cost of the special equipment and supplies that are necessary, and the cost of the labor involved in the operation of the mating hives.

Queen mating colonies are of two kinds: those "made up" with brood, and those made up without brood. They require different make-up methods.

Mating Virgins From Colonies in Full Size Hives

• **Board Divide.** A beekeeper's need for a few queens for home use can be met economically by using a mating colony called a board "divide" (Fig. 59). Two or three combs of sealed brood with adhering bees, and one or two combs of honey, are transferred from a parent colony into a body set upon an inner cover that has the center hole closed on each side with wire cloth or on one side with wood or other solid substance, and has a small entrance in the rim at one end. Additional bees from the parent colony are shaken from one or two combs into the divide body because the foragers that were put into that body are oriented to the parent colony entrance and will return to it. The divide body and inner

Figure 59. Board divide.

cover, *with the entrance to the rear*, are placed on the top brood body of the parent colony. A ripe queen cell is attached to one of the brood combs. Two weeks later there should be a laying queen that can be used elsewhere. If the divide is maintained it will continue to yield laying queens as long as they are wanted.

• **Requeening Colonies.** Requeening colonies with ripe queen cells is successful and is economical when done early enough for the colony to build large foraging populations for a honey flow. Virgins are mated from full size hives to requeen those colonies. Colonies to be requeened must be dequeened before a cell or queen is placed in the colony. Finding a colony queen can be a laborious task, but there are ways to simplify the work. (See Chapter V).

• **Queens for Colonies that are Divided for Colony Increase.** Queens for colony "divisions" that are made for increase can be mated from the divisions after they are made up. An excluder is slipped between the brood bodies of a two-story colony. After four days the body without eggs is set on a bottom board and is given a ripe queen cell. Moving the new colony at least temporally to another apiary will avoid loss of foragers that would return to the hive on the old location.

The same procedure is used to make two or three new colonies from one. One to two combs of brood with bees from the hive body without eggs and one or two combs of honey, plus an

Figure 60. Standard 10-frame hive divided into two 5-frame nuclei. Robber screen on entrance of one nucleus. From *Contemporary Queen Rearing*, Dadant and Sons, Inc.

Figure 61. Single 5-frame standard frame nucleus. Feeder screen on top of body.

empty comb, constitute a suitable start for a new colony. These are put in an empty body that is on a bottom board. A ripe queen cell is fastened between the brood combs. The queen for each division emerges in her division and mates from it.

Use of Special Nuclei for Mating Queens

The most cost-efficient and satisfactory arrangement for mating hundreds of queens for general use, or for sale, is mating the virgins from small colonies. These colonies are called *nuclei*.

Large nuclei may have standard brood frames in divided, standard eight frame or ten frame hive bodies (Fig. 60), or in individual three or five frame special bodies (Fig. 61). Other large nuclei may have shallow frames in divided standard shallow bodies (Figs. 62, 63), or special frames in special bodies.

The smallest units (Figs. 64, 65) are individual nuclei and are called "baby" nuclei. The nucleus preferred by German beekeepers is made of foam plastic and has its own particular shape. Different types of nuclei require somewhat different management.

• *Individual Five-Frame Nuclei.* Individual five frame nuclei whether full depth or shallow are excellent for continuous mating of queens and as queen reservoirs in outapiaries. They are established with the common arrangement of one comb of sealed brood and adhering bees, one comb of honey, and one empty comb. Bees from one extra comb shaken into the nucleus adds to the likelihood of a continuing nucleus. A ripe queen cell is attached to the brood comb. If proper drones are plentiful in the originating apiary the nucleus can be left where it is made up, but to avoid return of foragers to the parent colonies it is better to move nuclei to an outapiary where the proper drones are abundant.

• **Divided standard hive body**. A nucleus that is often chosen for mating queens, and which is adaptable to the needs of the honey producer, is the standard hive body divided into two separate compartments (Fig. 60). The dividing partitions are generally made of wood, and they are stripped on the ends and bottom to hold them in place and to make the compartments *bee tight*. Some beekeepers fit these partitions into grooves cut into the end walls of the hive, and then strip the bottom. By closing the compartment tops with individual wood covers, or with canvas tacked to the partitions, one compartment can be opened at a time.

The entrances of the nuclei are situated at the opposite ends so that no two are on the same side of the hive. This arrangement helps the queens and bees mark their own entrance and reduces the danger of bees drifting from one nucleus to another, and of

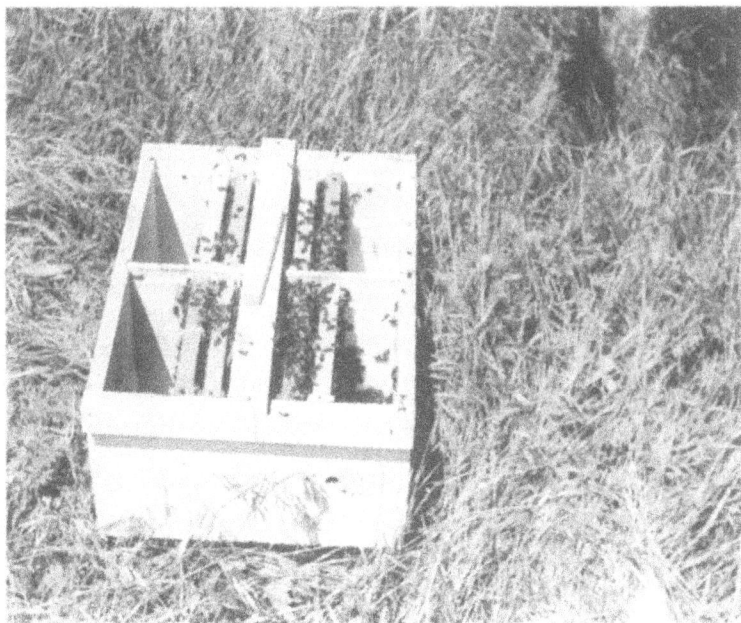

Figure 62. Standard 10-frame deep, or shallow, body divided into four compartments. Only one entrance on a side.

Figure 63. Shallow (5 11/16 inch) eight-frame body divided crosswise by a special double "division board" feeder to make two nuclei.

queens entering the wrong nucleus upon return from an orientation or mating flight. The frames, interchangeable with regular colonies, facilitate the addition or removal of bees, honey, and brood to keep the nuclei in good condition.

Each nucleus is made up with a comb of honey, an empty comb, a comb with sealed brood, and bees taken from populous colonies. To prevent forager bees from drifting back to their parent colonies, the nuclei are made up in one yard and then moved to another that is populated with drones of good stock. The entrances can be closed with green grass while the nuclei are being formed, but a robber screen placed over the entrance to completely restrict passage until nightfall is preferable because ventilation is not reduced. The robber screen can be left on the entrance permanently to protect the nucleus from robber bees. If that is done, a small entrance to one side of the screen or a small entrance through the screen at one side is necessary to help the queen on her returning from her orientation or mating flight to find an odor-marked entrance into the hive. A 1/4 inch space between the robber screen and the hive wall along the top of the screen forms a main hive entrance which is ignored by robber bees.

By assembling combs of sealed brood and bees from one set of colonies and extra bees from another lot if needed, nuclei can

Chapter IV

Figure 64. Typical baby nucleus, crumpled pigeon wire in can feeder. From *Contemporary Queen Rearing*, Dadant and Sons Inc.

be established rapidly. Each nucleus is given a comb of honey, one of brood, an empty comb, and about a pound of bees. A ripe queen cell is given to the nucleus just before the bees are put in. The brood will prevent the bees from absconding, and the emerging bees will maintain the strength of the colony until it can rear brood of its own. It is not recommended to establish this nucleus with less than two combs. While it is generally advantageous to use brood in the formation of nuclei, it is not essential, and when large numbers of nuclei are formed it may be impractical. *Nuclei formed without brood should remain closed for three days in a cool dark place before they are put on location* to allow the bees to establish a coherent colony with a virgin queen. During confinement the virgin emerges, and the bees blend to form a colony that renovates the combs and accepts the emerged virgin. When combs of honey are not readily available, division-board feeders are substituted for one of the frames in nuclei having three or more combs. Feeder screens between the body and the cover (Fig. 61) of single five-frame nuclei are excellent when feeding candy or pollen patties. Entrance feeders can be used, or sugar syrup sprayed into combs, but these may encourage robbing unless the entrance is reduced, or robber screens are used.

• **Baby Nuclei.** Small mating hives with frames approximately four-by-five inches are referred to as *baby nuclei*. They vary con-

Figure 65. Three-frame "expanded" baby nucleus. Crumpled pigeon wire in rectangular can feeder.

siderably in construction and size, but the typical baby nucleus (similar to the one shown in Fig. 64), is a box measuring five and three-quarter inches deep, four and three-quarter inches wide and five and three-quarter inches long on the inside, and with walls three-eighths inch thick. The box holds three four-by-five inch frames and a feeder. The entrance is usually a one-half inch hole in the front wall near the alighting board. The entrance can be closed with a metal strip, and the queen can be confined by a piece of queen excluder, which is seldom used. The cover fits down over all sides or simply over the ends. A one-inch screened opening in the back allows ventilation. In most situations the small entrance allows the bees to defend the little colony from robber bees. If robbing is *severe* a robber screen will protect the nucleus. A small entry hole in the robber screen, *offset* from the nucleus entry hole, is necessary to give an odor-marked access to the nucleus for the queen.

An expanded version of the baby nucleus (Fig. 65) is more adaptable to varying environmental situations than the smallest baby nucleus. The success of baby nuclei depends to a great extent on weather conditions and on how they are managed. Commercial queen breeders have used them for years because they are relatively inexpensive to make and to operate. They are strictly early season, short-time mating hives because they are too small to produce more than a small colony. Queens from three to five successive grafts can, as a rule, be mated from these nuclei before they are discontinued for the year.

For the first round of queens, the baby nuclei are stocked inside a building which is kept cool, by an air-conditioning unit if necessary. A sufficiency of bees is shaken from regular colonies in outyards into a ventilated bulk bee box (Fig. 66 a,b), that has a capacity for fifteen to twenty pounds of bees, and is provided with a frame of honey.

The baby nuclei boxes to be established are lined up in rows. A frame is removed from each nucleus and the feeder tins are filled with sugar syrup. Before the bees are added, a ripe queen cell is pressed by its base into one of the combs remaining in the nucleus. The bees in the bulk bee box are sprayed lightly with thin syrup, or with water, and a full half-pint of bees is dipped from the bulk box for each nucleus and is poured into the nucleus. The nucleus combs are replaced and let settle by their own weight before the nucleus is closed. The nuclei are left in the cool, dark room for three days and are then moved to their locations. The queens will have emerged by then from their cells and each nucleus will be a miniature colony. The baby nuclei are located in apiaries with partial shade and abundant landmarks, (Fig. 67). They are placed with two or more nuclei nearly together, and with the entrances in different directions. Sometimes nuclei are placed on platforms for ease in handling, or are arranged in circular fashion so the beekeeper can manipulate each from the circle center.

During favorable mating-weather conditions, the queens will begin laying in about ten to fourteen days. Approximately twelve to seventeen days after the cells are placed in nuclei each nucleus is examined and the feeder refilled with light sugar syrup. If the queen is laying, she is caged either for use or for shipment and is removed from the nucleus, or the nucleus is marked to indicate its condition and the condition of the queen. Before leaving the apiary, a ripe queen cell is put in nuclei from which queens were taken. Brood from other nuclei is added to those that were queenless because the bees are too old to support the nucleus until brood from the new queen emerges. A queen cell is then given to it. Lack of a queen is known as a "miss".

Usually the queens are left in the nuclei until they have laid in one or in two combs to be sure that the queens are laying well,

Figure 66 a and b. A bulk bee box. From *Contemporary Queen Rearing*, Dadant and Sons, Inc.

Chapter IV

and to leave brood for nucleus maintenance. A ripe queen cell may be put in at the same time a queen is caged or on the following day. If a queen is not laying by the seventeenth day, some queen breeders will destroy her and give the nucleus another queen cell if the strength of the nucleus justifies it. Otherwise, the nucleus is taken in and reconditioned. The weather should be taken into consideration in determining whether mating delay was caused by weather. Mating delayed by weather for two weeks may not prevent the queen from mating on the first favorable flying day.

• **Other Types of Nuclei for Mating the Queen.** Other types of nuclei are also used for mating queens. Their operation varies according to the nucleus size, but the same principles govern the use of each type.

Nuclei made from shallow (five and eleven-sixteenth inch) eight-frame extracting supers are preferred by some queen breeders. (By definition a shallow body is any hive body less than standard full depth). The shallow bodies are divided into two or four compartments and are fitted with a bottom board. They can be divided lengthwise into two compartments to accommodate the standard shallow extracting frame, or crosswise to form four compartments that require special frames. One type of nucleus made from the eight-frame shallow extracting super has a double division-board-type feeder separating the body crosswise into two compartments (Fig. 63). Each compartment of this nucleus will take five combs. This body is sometimes divided into four nuclei. The entrance of each nucleus is on a different side. Queens are caught most rapidly when several nuclei can be worked before it is necessary to stand and move on. Sixteen of these nuclei can be loosely clustered in four bodies in a circle.

When queen rearing is finished for the season, the division boards are removed, and bodies with the combs are stacked on bottom boards, each stack with bees and one queen. The bees fill combs with honey for the next year. The combs in these stacks will be cared for by the bees until needed.

A standard ten-frame hive, or shallow, is sometimes divided into four compartments, one in each corner, by division boards running lengthwise and crosswise (Fig. 62). Each compartment has a division-board type feeder, and the entrance of each compartment is located on a different side of the hive. Frames for this nucleus are made from standard Langstroth end bars and special-length top and bottom bars. Those who use this type nucleus cite the advantage that one can see the entire surface of the small comb at a glance, thus aiding in finding the queen. This nucleus combines the advantages of the double divided hive nucleus and the larger baby nucleus.

Some Shortcuts in Establishing Nuclei

Nuclei can also be made up from storage by setting the prepared bodies over the broodnest of regular colonies for a period. The colony queen will lay in the combs and some honey will be stored.

Some beekeepers make up nuclei by arranging small combs in special holders in hive bodies that are placed on top of ventilated cages of bees. If they are left overnight, the frames can be

Figure 67. Nucleus apiary in partial shade with abundant landmarks.

separated and put into nuclei the following morning. When establishing nuclei that have Langstroth frames, brood and honey can be assembled from different colonies and stacked above a queen excluder on a few strong colonies. Within a short time, all of the combs will be covered with bees and can be set into the nucleus hives. Another way of accomplishing a similar result is to pile hive bodies of brood on regulation bottom boards and add about five pounds of bees for each hive body of combs.

It is apparent that there are numerous ways to establish nuclei. A beekeeper is likely to use more than one.

Arrangement of Nuclei

The established nuclei are taken to mating "yards" (Fig. 67) where they are placed on the ground, or on benches at a height more convenient for the beekeeper handling them.. Nuclei are arranged in pairs, or in irregular groupings, with the entrances in different directions, and with sufficient space between the different groups to prevent drifting of bees or queens. Partial shade is needed for baby nuclei. Protection from ant attacks is necessary.

Maintenance of Nuclei

The time required to take care of nuclei depends considerably on the type of nucleus, the purpose for which it is being used, and floral conditions. At each examination, all *necessary* work should be done. None should be left to some future date.

Some queen rearers prefer nuclei made with standard frames, or even shallow extracting frames, that are interchangeable with those in honey producing colonies and therefor easier to manage. Divided hives containing three standard full depth frames are self-supporting in an average location; but if queens are to be held in them during the main honey flow, they are checked at weekly intervals. Such nuclei are ideal to draw combs from foundation. They are a good source from which to make increase during the season and they can be wintered in the milder climates and divided in the spring to make more nuclei.

The small frames of baby nuclei can be assembled in racks in standard equipment at certain times of the year and placed on colonies where they will be filled with honey or brood for future needs.

During the summer months, when the weather is hot, it is more difficult to maintain baby nuclei than larger nuclei, because the number of bees is not large enough to cope with the ventilating needs and other requirements of the little colonies. Queen breeders usually discontinue them entirely during the hot season. Partial shade is desirable for all small colonies. Use of double covers to provide insulation from the heat helps the colonies keep their hives cool enough to permit normal activities.

It is hard to say just how large or small a nucleus should be. Requirements vary greatly with environmental conditions. A nucleus containing one pound of bees in a space large enough to permit it to expand without clustering out during warm weather is usually suitable; it can keep the hive cool in partial shade and also keep in warm on the cooler days and nights to early spring.

Feeding Nuclei

The best food for nuclei is honey stored in the comb. A nucleus should have ample food stores at all times. It is difficult to rear queens when the mating colonies have meager supplies of honey and pollen and the nectar and pollen flows are very light. During seasons when nectar is scarce or when the nucleus is too small to be self-supporting, the danger of robbing is increased somewhat by feeding sugar syrup, though it must be done. When combs of honey are not available, a heavy sugar syrup (2 parts sugar/1 part water) fed in a feeder enclosed in the nucleus is the best substitute.

In moist climates, a nucleus containing a half pound or more of bees will survive on dry granulated or powered sugar, especially finely granulated sugar. When water is readily available, the nucleus bees will dissolve the sugar and build comb, and even store the sugar in liquid form in the comb. This method of feeding incites robbing less than any other method when artificial stores have to be supplied. The granulated sugar can be placed in the feeders or poured on the bottom board at the back of the hive. The bees will carry some of the sugar out of the hive, but will use enough of it to maintain the colony. A powered dry sugar called Drivert® is the best of the dry sugars for feeding nuclei.

Methods of supplying sugar syrup to nuclei differ with the size of the queen rearing operation. Containers for handling the syrup vary from sprinkler cans with the sprinkler head removed, to five-gallon pressure sprayers, and even to spray tanks holding 100 gallons or more of syrup, and usually equipped with hoses and cutoffs. The last type is used in very large mating yards, where several thousand nuclei are to be fed. Two beekeepers apply the feed as a third drives through the apiary.

Mating-Hive Records

Operational records are a necessity whether few nuclei or thousands are operated. Records should be simple and easily kept.

Nuclei can be numbered and records kept in a notebook with a page devoted to each nucleus; or the records can be written on the top or side of each box with a heavy crayon. Other types of records have been devised that are preferred by some beekeepers. One consists of a wooden block that may be colored differently on the different sides, which is shifted to various positions on top of the nucleus to indicate different conditions of the nucleus. Some beekeepers utilize charts and pointers on the nucleus.

The date that the cell is put into the nucleus is recorded. When the queen is removed for use or shipment a line is drawn through this entry. If the queen disappeared from the nucleus, the "skip" is indicated by circling the date, and brood is given the nucleus from another nucleus. If two successive skips occur in a nucleus the bees have become quite old and sealed or emerging brood is given from another nucleus, or the nucleus is taken to the shop to be restocked.

When more than a few nuclei are operated, record keeping is greatly simplified and time spent in searching for laying queens is reduced if the nuclei are established in small yards, or in blocks and "sites" in larger yards similar, in a way, to arranging cell builders in rows. A block may consist of enough nuclei to take care of one day's operation. An entire block is established at one time and the block is worked at one time. If the blocks are numbered, they are referred to as easily as to single nuclei. Written top records are often kept on the individual nuclei of a block.

At the time queens are caught from nuclei, some nuclei will need feed, brood, bees, or other attention. These needs should be met after the queens are caught and before leaving the yard. The operation is more efficient if the various needs of individual nuclei are signaled when they are first discovered. Rocks or clods of dirt placed on the top serve very well as temporary indicators of the nucleus' condition and needs and they can be discarded when the nucleus has been reconditioned and before the apiary is left.

Robbing

Robbing is one of the most serious problems confronting the queen producer. It is especially difficult to control when baby nuclei are operated in areas where the spring nectar flow is very light or erratic.

The best robbing control, all beekeepers agree, is prevention. Unless it is absolutely necessary, nuclei should not be worked on days when robbing is almost certain to occur. A minimum of smoke is sufficient when examining nuclei, and the nuclei should not be kept open longer than necessary. Honey, syrup, or bits of comb left on the ground or otherwise exposed to searching bees incites robbing. Feeding should be done late in the afternoon, if practical, and it is preferable to feed by giving combs of honey, which should be placed in the nucleus away from the entrance, or by sugar syrup in a division-board type feeder similarly located.

It is sometimes necessary to catch queens, put out cells, or for other reasons open the nuclei even when robbers are menacing. The open nucleus may be protected by a damp sack laid across the top while a frame is being examined, and after the nucleus is closed grass can be piled loosely over the entrance. If robbers seriously attack the nucleus they may be discouraged by laying a wet sack over the top of the nucleus. Contracting the entrance of a nucleus, that has a separate screened ventilation opening, so that only two or three bees can pass through the entrance at a time will allow the nucleus to defend itself. When the entrance is the only source of ventilation the robbers attempting to enter the nucleus, and the colony bees attempting to get out to attack the robber bees, may mass so tightly at the restricted entrance that the bees smother within. When this is a potential danger, a special entrance screen can be used to restrict the entrance while allowing almost full ventilation. This device, the *robber screen*, (Figs. 60, 63, see Remarks) that is constructed like a small moving screen, permits the defending bees to come out of the entrance and crawl over the front of the nucleus while protected by the screen. A quarter- to three-eighths inch open space along the top between the screen and the hive body becomes the main entrance to the nucleus. A small open passage into the nucleus at one side of the screen frame, or a small horizontal slit in a lower corner of the screen, allows a queen returning from an orientation or mating flight to find an odorous opening into the nucleus. The robbers try to gain entrance through the screen and the small nucleus opening, but the bees of the nucleus guard the side opening and repel the robber bees. The robbers ignore the opening along the top of the screen, while the bees of the colony use it freely.

THE CARE AND USE OF QUEENS

General Practices

Since the queen is the most important individual in a honey bee colony, the frames, excluders, and bodies must be handled in a manner that avoids injuring her. In handling frames, one must be alert to the possibility of crushing the queen between the shoulders of the frames, between burr combs or uneven surfaces of the combs, or against the sides of the hives. Pushing the combs away from a side comb creates a larger space for frame withdrawal.

Queens in cages should not be exposed longer than briefly to the direct rays of a hot sun. This applies to queen cells as well. Prolonged exposure to cold may damage adult queens, and the wings of pupae in queen cells can be damaged so they do not expand properly. Colony examinations should be avoided when colonies have virgin or newly introduced queens or queens that have just mated. Such queens, and the bees of the colony, are easily excited, and the disturbance of jarring or of opening a hive sometimes causes the workers to "ball" the queen. When this occurs, one or more bees will attack the queen and attempt to sting her. Other bees will join in the affair, biting and pulling at her legs and wings. Soon a ball of bees will be formed tightly around the queen. Queens frequently come out of the experience with a paralyzed leg or some other injury, or may be killed, although occasionally a queen escapes bodily harm. Should balling occur while working a hive, a few puffs of *cool, dense* smoke will disperse the bees; hot smoke will cause the bees to sting the queen.

Finding Queens

In the routine manipulation of colonies in full size hives much time is lost in unnecessary searches for queens. Finding the queen is not necessary in most beekeeping operations. One can evaluate a queen better by examining her brood and its distribution in the hive than by looking at her on the combs. At certain times, however, it is necessary to find the queen. Finding a queen in a colony can be a laborious task, and a few suggestions might be helpful.

Laying Queens

Using only enough smoke to control the bees, a side comb of a broodnest body is withdrawn to provide space for handling the other frames. The search for the queen in a one-story hive may begin with a middle comb of the brood body and continued alternately toward the sides. Queens are photophobic in the hive and tend to shun light. As a frame is removed from the broodnest the queen moves away from the light of the vacant space. A quick glance down the sides of the combs on each side of the vacant space as the combs are withdrawn often reveals the queen. Before a hive is opened, the queen is usually on the comb area where she is depositing eggs. With little or no disturbance she is likely to remain there, or nearby. Much smoke or other disturbance will cause her to run so she may be anywhere in the hive.

Between bursts of egg laying, the queen is often surrounded by a broken circle of bees facing her, and as she moves across the comb the bees give way from in front of her (Fig. 68). This arrangement of bees frequently strikes the eye about as quickly as the sight of the queen herself, and it aids in locating the queen on the comb.

Figure 68. Catching laying queen. Bring fingers from behind along sides of queen. Grasp wings on both sides without pressing on queen, or grasp the queen's thorax by the sides.

Various manipulations in beekeeping are facilitated by confining the queen to one or two brood chambers by a queen excluder. The search for the queen in the brood bodies of these colonies starts by setting the honey supers, that are above the queen excluder, off onto a "superhorse" (see Remarks) or upturned hive cover. The excluder is removed and the underside examined for the queen before reversing the excluder and placing it on top of the supers. Then, using no more smoke than is necessary to control the bees, the top brood body is set on the excluder that covers the supers, or on an upturned hive cover, or on a superhorse, and is examined. The lower body is then examined if the queen is not found in the top body. Routinely inserting excluders between the two brood bodies of two-body brood nest hives at least four days before searching for the queen is a time saver. The presence of eggs after four days will indicate the location of the queen, and the search is limited to one brood nest chamber of each hive.

When the queen cannot be found by these direct methods, although eggs indicate that she is present, the top brood body is removed and set on an *empty* full-depth or shallow hive body that is separated by an excluder from a lower body that has only two combs. The bees are shaken from the brood combs into the empty body. With an occasional puff of smoke, the bees move through the queen excluder and into the lower body. A glance now and then at the bees on the queen excluder will locate the queen if she were shaken along with the bees. If the queen is not found, the process is repeated with the second brood body.

An effective alternate method, is to blow cool, dense smoke into the hive entrance to drive the bees up into a "cluster box" that is positioned on top of the brood nest. After about three to five minutes the cluster box is removed from the hive and the bees are jarred from the cluster box into an empty body with an excluder beneath it, that is placed over the brood nest. The workers will pass down through the excluder and the queen should be found on the excluder. A "cluster box" (Laidlaw 1979) is a shallow body screened on top that permits air and smoke to escape from the top of the hive. It is fitted with lengthwise or crosswise wooden slats that form perpendicular baffles to which the bees cling after they are driven up from the brood nest by the smoke. Though not necessary, the upper edges of the slats may be scalloped to permit passage of bees between slats. Using three or four cluster boxes in tandem and leapfrogging colonies as the queens are caught reduces the time involved in dequeening.

Another efficient variation is to place a double rim "queen screen" (which is two 1 1/2 inch depth rims fastened together with an excluder between them on top of the booodnest. As the hive is smoked at the entrance the bees run upward and pass through the excluder but the queen is caught beneath it. After the bees are smoked up, reverse the rim and lay it on top of the brood-nest body. The workers go down through the excluder exposing the queen above the excluder. If needed, a fine spray of water will reduce flight of bees above the excluder.

A "shakerbox"(Laidlaw 1979) that is widely used in shipping package bees, is invaluable for finding queens by screening the bees through an excluder when the bees are needed for various purposes.

Virgin Queens

The absence of eggs and the presence of the telltale mark of a queen cell from which a queen has recently emerged, or remnants of queen cells that have been torn down, indicate that a virgin queen is probably present. Virgin queens are treated differently by the bees than are laying queens and they do not have a retinue of bees around them. They have a smaller abdomen, and are more prone to run and hide under bees than are laying queens. They may even try to fly if an attempt to catch them is unsuccessful. If careful examination of the combs fails to locate a virgin queen, forcing the bees to go through a queen excluder is probably the quickest way to find her in a populous colony. The queen can be caught on an excluder that is placed under an empty body, or on an excluder that is placed on the entrance. The bees are shaken from the combs into the empty body or onto the ground in front of the entrance. The queen will be found on the excluder as she tries to pass through it. A temporary ramp, such as the hive cover leading up to the entrance, will hasten the bees' entering the hive.

Virgin queens, swarm queens, or laying queens that have ceased laying, may fly if released from a cage or disturbed on a comb. These queens usually return after a brief flight to the place from which she flew, if the place remains unchanged. She will return to her cage, or comb, or even to the beekeeper's hand if she flew from the hand and if the hand is still in place.

Queen rearing nuclei are just small colonies used in mating queens. They have fewer frames and bees than the larger hives, and their queens are much easier to find. A glance down between the combs of a nucleus will indicate where the brood is likely to be, and the comb with the greatest number of bees should be examined first. Very little, if any, smoke need be used in examining nuclei, and the queen is likely to be on the combs. If she cannot be found there, then all frames must be removed and the search extended to the sides and bottom of the box. Before caging the queen for use or shipment, it should be determined that she is laying normally.

Queen Cage Candy

Queen cage candy is useful whenever queens, worker bees, or drones are confined. In the early commercial production of queen bees, considerable difficulty was experienced in securing a food that will sustain the queens without ill effects during the time they are confined to their cages in shipment. The best food for bees is honey in the comb, and this was tried first. Its use was soon rejected because honey in the small combs was prone to leak from the package, or to smear the bees in the package. Then a combination of honey and "pounded sugar" was used, mixed together until it formed a firm candy that would hold its shape in hot weather. This proved to be a satisfactory shipping food, but honey is no longer used as a constituent of queen-cage candy because of the possible danger of spreading American foulbrood with any honey that came from an infected colony. (Some beekeepers still use this type candy in their own cages for use in their own bees, selecting a well ripened honey which they know was produced in an area entirely free of foulbrood.)

The candy that is now used universally for commercial shipment of queens is the "Good" candy. It is made by mixing invert sugar syrup and powdered sugar in approximately a 1:3 ratio. The powdered sugar is stirred into the sugar syrup. When the mixture becomes too thick to stir, it is kneaded on a slab with powdered sugar. More powdered sugar is added as the candy is kneaded until a firm candy is formed that will not run. Much of the candy is currently made with a commercial liquid inverted sugar known as Nulomoline™ that is produced without the use of acid. Some beekeepers add a small amount of glycerin to the syrup which tends to prevent the candy from drying. Queen cage candy should be left to stand for some hours in the sun or warm area before using, so the sugar and syrup are thoroughly mixed, and so the candy's firmness can be checked. It should not be hard and dry, nor soft and sticky. In hot weather, more powdered sugar is required than in cooler seasons. If properly made, the candy can be stored in an air-tight container for a considerable time.

High fructose corn syrup has been used satisfactorily in making queen cage candy and in shipping package bees. Drivert® a partially inverted powdered sugar is commonly substituted for powdered sugar.

Queens and bees can be shipped through the mails for a total period of ten days when supplied with queen cage candy. They will not survive a much longer period unless they have access to water to offset the dehydrating effects of the sugars. The small amount of starch in powdered sugar has little, if any, ill effects on the workers or queens. The main ingredient that is lacking in queen cage candy, as made at present, is an adequate amount of moisture.

Catching Queens

It is usual to catch a queen in order to cage her or to clip and mark her. She should be caught from behind by slipping the thumb and index finger down over the side of her thorax and abdomen and grasping her wings of both sides without pressing on the abdomen. Or, her thorax is grasped between the thumb and forefinger and she is held between them.

Clipping and Marking Queens

When the queen is to be clipped and marked, her thorax should be held between the thumb and forefinger (Fig. 69). It is best to clip off less than half of *either* the right or left wings, leaving the other pair to pick her up with in the future. Clipping *either* the right or left wings in even years and the opposite wings in odd years enables the beekeeper to know the age of his queens.

Queens are marked (Fig 70) to help detect them in the colony or to convey information about their age or breeding. A bright spot of color, such as red, green, or yellow, on the top of her thorax makes a queen stand out on a comb rather conspicuously. This is especially true of the dark races of bees. Fingernail polish,

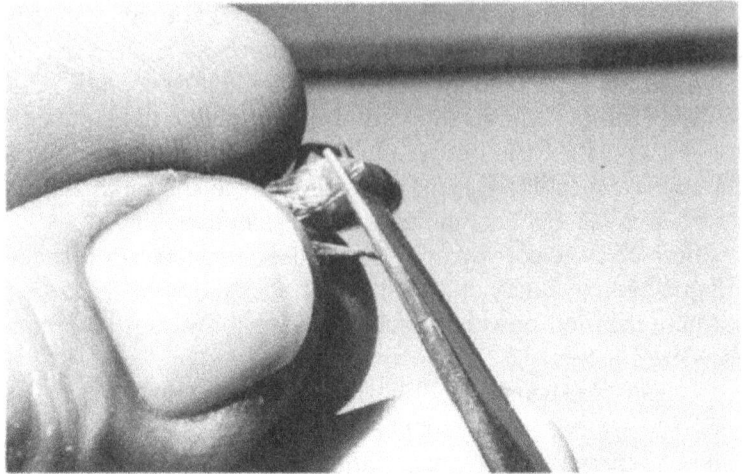

Figure 69. Clipping a queen. Hold queen by thorax with abdomen extending from end of fingers.

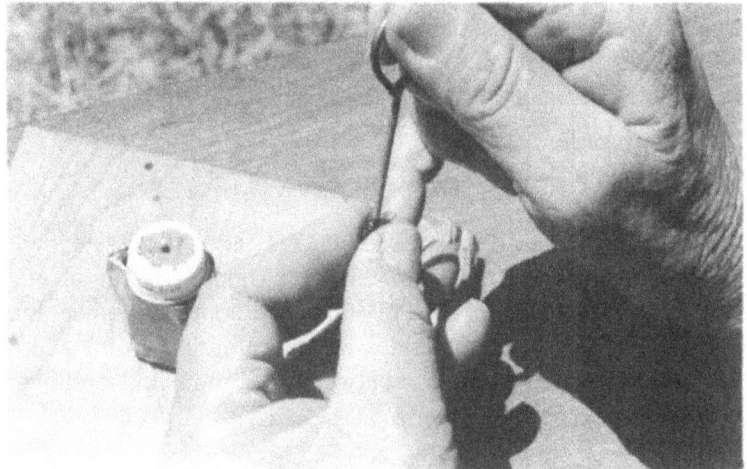

Figure 70. Marking queen. Use flat-end probe to make smooth round mark. A hole in top of paint bottle scrapes paint from side of marking probe as probe is withdrawn. From *Contemporary Queen Rearing*, Dadant and Sons, Inc.

a quick-drying lacquer like automobile touch-up lacquer, or a dry pigment suspended in alcohol and shellac, or celluloid dissolved in acetone, make satisfactory marking fluids. Numbered or plain disks that cover the back of the thorax are available in several colors to apply to the thorax of queens, and are carried by bee supply firms.

The International Color Code for marking queens employs five colors: *white* for years ending in 1 or 6; *yellow* for years ending in 2 or 7; *red* for years ending in 3 or 8; *green* for years ending in 4 or 9; *blue* for years ending in 5 or 0. Colors applied to different portions of the queen's thorax or second segment of her abdomen can indicate breeding and origin, and two or more colors can be used for detailed information.

Inexperienced beekeepers will develop confidence by marking and clipping drones. Pigments should be applied *carefully* to prevent any fluid from getting on the neck of the queen or covering any of the spiracles of her thorax or abdomen. If properly applied, a good lacquer will not injure the queen and will serve as a distinguishing mark for as long as she lives. Pigments are sometimes suspended in fluids that excite the bees when the queen is returned to the comb. Holding the queen a few seconds until the fluid dries and some of the odor is dissipated minimizes this problem. The vapor accompanying any of the marking fluids should be nearly gone before the queen is returned to her colony. Considerable time is saved in finding queens of the dark races in colonies if they are allowed to emerge in nursery cages in their nuclei or hives and are marked before they are released.

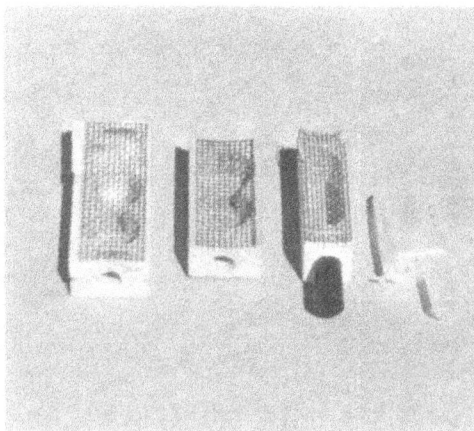

Figure 71. Queen shipping cages. Benton three-hole, two-hole, California minicage and JZ^s BZ^s cages.

Caging and Shipping Queens

Queens must be caged when they are transferred, transported, introduced to colonies, or stored. Cages of various designs have been used over the years. Most were satisfactory for the use to which they were put. The shipping cage invented by Frank Benton has been the standard queen shipping cage for about a century. It has also been used as an introducing cage equally as long. Smaller cages and plastic cages (Fig. 71) are also used now for shipping queens individually or in groups of caged queens. The "battery box" is an efficient device for shipping groups of queens, caged without attendants, to one destination.

A queen is caged by inserting her head and forelegs into the entrance opening of the cage and giving her a gentle push with the middle finger before entirely releasing her wings (Fig 72). A queen can also be caged safely by "herding" her into the end hole of a cage placed in front of her while she is crawling on the comb (Fig. 73). The forefinger and thumb of one hand direct her movement while the other hand holds the cage in place.

A worker that is to accompany a queen that is to be mailed is easily caught when she has her head inserted in a comb cell with her wings protruding from the cell. Her wings are grasped so her head is toward the beekeeper's wrist. The same orientation of the

Figure 72. Caging queen in shipping cage. Place queen's head in the entry hole. Position fingers to prevent the queen from backing, and let her run into the cage. From *Contemporary Queen Rearing*, Dadant and Sons, Inc.

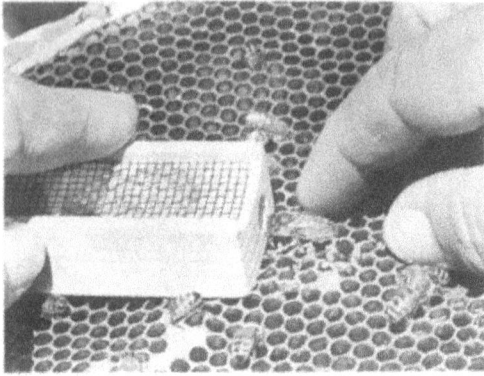

Figure 73. Herding a queen into a cage. Guide the queen as she crawls on the comb to the entrance hole. From *Contemporary Queen Rearing*, Dadant and Sons, Inc.

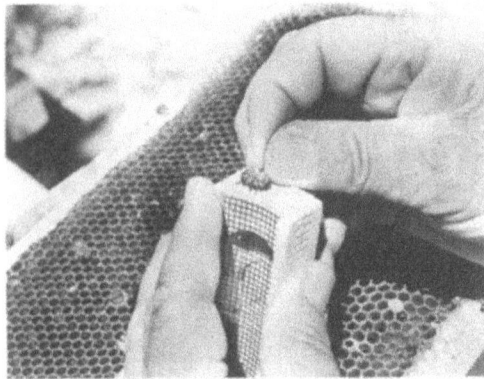

Figure 74. Caging queen attendants. Cage workers that are taking honey from cells. Grab them by both pairs of wings. Stick the worker's head into the cage entrance hole and release the wings From *Contemporary Queen Rearing*, Dadant and Sons, Inc.

worker applies when she is caught on the face of a comb. The worker's head is inserted into the entry hole of the shipping cage, as shown in Fig. 74, and, as the wings are released, the worker crawls into the cage. The beekeeper receives fewer stings in the thumb when workers are caged this way and the cages are rapidly stocked with queen attendants.

Another way to stock numerous mailing cages of queens with worker attendants is to shake one or two pounds of bees from open brood into a hive body and spray the bees lightly with thin sugar syrup. Ten to fifteen minutes later the bees are full of syrup, and they will not attack the queens. The cages with laying queens are set on end, with the cage entry holes open, over excludered holes in a thin board that covers the hive body with the intended escorts. In a few minutes enough bees enter the cages to care for the queens.

A large queen rearing operation in Kona, Hawaii developed a very efficient way to introduce worker attendants into cages with queens with a vacuum device. This is proving to be an economical labor saving technique when many thousand queens are shipped each year.

The Benton 3-hole cage (Fig. 71) is the venerable, highly successful, common cage for shipping one or more queens by mail. The paraffined chamber of the cage is provisioned with queen-cage candy, and the candy is covered with a square of wax paper to retard dehydration of the candy. The outer end of the end hole that extends into the candy compartment is stoppered with cork or thin cardboard. The queen is caged (Figs. 72 and 73), and six to ten attendant bees, taken from the same hive, are added, as is shown in Fig. 74.

One, or several queens that are caged in seperate cages with attendants and are enclosed in a manila envelope that has several small holes for ventilation, ship satisfactorily in the mail. Caged queens that are crated or bundled ship equally well.

Bees for combless packages are shaken from several colonies and mixed as they are caged. The packages always include a young laying queen that was taken from a mating nucleus. The queen, caged alone in a two-hole cage without candy, is suspended in the package, and is accepted by the bees in the package during transit. The loss of queens in shipping or after hiving is rare.

Instrumentally inseminated queens from nursery colonies that

Chapter V

are shipped as components of combless packages are introduced to the package bees enroute (Nelson and Laidlaw 1988).

For decades, groups of queens caged with or without candy have been shipped successfully in a single combless package. Roy Stanley Weaver, Jr. developed this into a popular method of shipping as many as 150 caged queens to one address by a means of a wooden or cardboard crate-like "battery" box (Fig. 75), probably so named because of Weaver's artillery officer experience. The battery box is solid on top and bottom and is screened on the sides and sometimes only on the ends. The queens are caged without attendants and with no candy in the queen cages. The queen cages are set on end in rows in racks in the battery box, and candy is put on the bottom of the box. Approximately one-half pound of bees is shaken from the broodnest of a good colony into the battery box and the cover is fastened on the box.

Figure 75. Battery box that was invented by Roy Stanley Weaver Jr. for shipping a collection of queens that are caged individually without attendants or food in their cages. The queen cages are placed back to back and on end in lengthwise racks. Candy is placed on the battery box floor. About one-half pound of bees from a hive body above an excluder with emerging brood is poured into the battery box before the top is put on. This is a modification of the practice of shippping extra queens in combless packages. Battery box supplied by Gus Rouse, Kona Queen Company, Hawaii.

Queen Pheromones

Soon after a queen is removed from her colony the bees exhibit considerable "distress" and a colony-wide buzzing may be audible. Unless a queen is restored to the colony within a short time queen cells will be started around young larvae. Butler (1954) determined that a queen produces a material, which he called "queen substance, "that enables the bees to sense her presence in the colony." A failing queen apparently is unable to produce a sufficient amount of the substance to prevent the bees from starting supersedure queen cells. Butler speculated that the substance is acquired from the queen's body by bees that add it to their food and distribute it mouth-to-mouth to other bees in the colony. This substance which generates an action in other individuals of the same species is known as a *pheromone*.

Subsequently, the source of the queen substance was discovered by Butler and Simpson (1958) to be largely in the mandibular gland of the queen. They thought that the queen distributed the substance over her body when she cleaned herself, with the bees getting it as they stroked the queen with their tongues. But this method of dissemination does not adequately explain how the absence of the queen is detected so quickly by the colony after her removal. Beekeepers know that bees are attracted by the odor of crushed queens, or to cages in which queens have been held, or even by the odor of queens in queen cages before any contact has occurred.

Gary (1960, 1961) found that queens became less attractive to workers when the mandibular glands were removed. He also reported that worker bees, on which contents of the mandibular glands of a queen were placed, incited antagonism toward them by other workers in the same colony. Bees which came in direct contact with these tinged bees were attacked by still others, just as queens are attacked in a strange colony. The chemical composition of the substance has been determined to be 9-oxodec-2-enoic acid; it has been synthesized by Callow and Johnston (1960) and is currently sold as a component of Fruitboost® by Phero Tech, Inc.

Pheromones are pervasive, and have a complicated role in honey bee activities, among which is the acceptance and recognition of a queen as the colony queen. Habituation has a place in the introduction of queens also, as is shown by a strange queen being accepted by a colony after a period of confinement among the colony bees and in contact with them.

Introduction of Queens

A good honey flow facilitates most beekeeping manipulations, including the introduction of queens. If a good honey flow is not in progress, the colonies to be requeened should be fed a light or medium sugar syrup continuously until the introduced queens are laying well. A light sugar syrup more nearly resembles nectar in sugar content than do heavier syrups, and it is the *incoming* nectar that modulates the mood of the colony.

The most basic element of introducing a queen into a colony is that the colony be queenless. Any queen cells that may have been started during a period of queenlessness must be eliminated, and any emerged virgins removed. The second element is that after the new queen has been put into the hive the colony must not be disturbed until she has had time to establish a brood nest. A safe margin is about ten days. A queenless colony is more excitable than a queenright colony and does not lose all of its nervousness until after it has brood. A newly introduced young queen is also somewhat excitable until she has had time to lay normally without disturbance. The third consideration of importance in introducing a queen is that she be placed in the brood nest of the colony in a manner that will give her protection until the bees become accustomed to her presence and accept her as their own. Except placing a queen on a comb of emerging bees, queen introduction methods are not infallible.

Ready-to-emerge cells placed in the hive are the easiest and the surest way to introduce virgin queens into mating nuclei or "divides", or even to full colonies.

Success is likely if the queen is introduced into a broodless hive with young bees. A queen is usually accepted by a colony that has mostly sealed or emerging brood. Queens caged and placed among bees shaken from two or three broodnest combs of good colonies and confined for a day or two with food and ventilation before installation in a hive are rarely lost.

Completely disorganized queenless bees seldom fail to accept a laying queen. Bees dequeened and then shaken from their hive onto the ground in front of the entrance will often accept a laying queen that is dropped among them as they move back into their hive. In all cases the bees receiving the queen must be fed.

As a rule, queens can be introduced successfully if the colony is made queenless and the queen is introduced by a cage method in the same operation. The best time to place the queen in the hive is soon after the bees miss their queen and are making a characteristic buzz .

Cage Methods of Introduction

Queens have been routinely shipped in Benton mailing cages for decades (Fig 71). This wooden, screen-covered cage has three communicating compartments, one of which the queen producer fills with queen-cage candy. In shipment, the queen is accompanied by seven to ten worker bee attendants. This cage may serve also as an introducing cage when the queen is received. The bees of the colony to be requeened may show animosity toward the strange worker bees, and the workers are sometimes removed from the cage before placing the queen and her cage in the receiving colony, although it is not absolutely necessary to do so. The removal of these bees should be done inside a room or other enclosure, because the queen is light enough to fly, and may escape if the workers are released in the open. If a queen does escape, lay the cage with some of the escorting workers in it close to the spot from which the queen took flight. Move nothing else, so the environment to which the queen oriented is unchanged. In nearly all cases the queen returns to the cage. The queen cage usually has a piece of thin cardboard over the candy hole, and it is best to remove any restriction over the candy opening before the cage is placed in the hive.

After the colony has been made queenless and the worker bees have been removed from the cage, the cage should be placed between two frames in the brood chamber with the wire side either up or down and the candy end of the cage *up* so that any bees dying in the cage will fall away from the candy and not clog the cage exit. The bees in the colony need access to the wire screen of the cage in order to become acquainted with the queen and to feed her through the wire. It will usually take the bees about two days to eat a hole through the candy that is large enough for the queen to walk through. The JZ^{s}BZ^s Queen Cage (Fig. 71) is also both a shipping cage and an introducing cage. It is plastic with vent spaces along the sides that give the bees access to the queen but are small enough to protect her from bees' biting her feet. Two end tubular openings, one shorter than the other, are filled with queen cage candy which is the food supply during transit, and when eaten by the receiving colony releases the queen. The candy in the shorter tube is consumed before that in the longer tube allowing a few of the receiving bees to enter the cage with the queen before she is released. A small wooden cage that was developed in California is being used to ship queens in "battery boxes" It fits more easily between frames than the Benton cage, and the shipping candy is contained in a one-inch segment of a 3/8 inch (internal dimension) plastic tube. This tube serves also as an introduction device.

The push-in cage, under which the queen is caged over empty brood cells, some emerging brood, and some open cells of honey, remains a successful device. This cage, too, may have a tube filled with candy to release the queen from the cage when the candy is eaten by the bees that are receiving the queen. A second, but shorter, tube filled with candy is sometimes made a part of the cage to permit bees to enter the cage with the queen before she is released. If this cage does not have a queen escape tube the queen is released manually after about four or five days. The queen will probably be laying. If bees are behaving antagonistically toward the queen by biting the cage wire, the queen is left caged until the animosity ceases. If the bees persist in biting the cage after four days the colony should be examined thoroughly for a virgin queen.

Colony acceptance of queens, including those received by mail, is enhanced when three to five *honey-gorged* workers of the receiving colony are caged with the queen before she is placed in the hive. The queen will be safe with the engorged bees, *but* she will be killed if a non-engorged bee is caged with her!

Direct Methods of Introduction
Several methods of requeening a hive without using cages are used successfully by some beekeepers.

• *The Spray Method.* In using the spray method, colonies are made queenless and in the same operation, or a half hour or even a day later, the sides and tops of the frames in the brood chamber are sprayed with a fine mist of light sugar syrup, which is also used to wet the queen in her cage. The queen is then released directly on top of the frames and is given another spraying of syrup as she walks down between the combs. The hive is closed. One little-liked feature of this method is that in cleaning the sugar syrup from the queen the bees sometimes remove some of her hair, thus making her look older than she actually is. The method should not be used when it will instigate robbing. This method of introducing queens is very successful with package bees because the queens are introduced to the bees in the package in transit.

• *The Smoke Method.* After the colony has been made queenless the new queen is run in at the entrance of the hive, followed by several generous puffs of cool, dense smoke. This method works best when the new queen is taken from a nucleus only a few minutes before she is introduced to her new colony. Feed the colony before smoking the queen in, and continue feeding until the queen is laying.

• *Transfer of Combs of Brood and Bees.* Still another method of introduction to the colonies to be requeened in the apiary is useful when queens are mated from standard frame nuclei in an apiary which also has established colonies in standard frame hives. A colony is made queenless, and the frames of brood and bees from a nucleus, with their laying queen on an inside comb, are set over into the queenless hive in exchange for a like number of combs. A spurt of light sugar syrup spray over the sides and tops of the frames from the nucleus diverts the attention of many of the bees so they fill themselves with the syrup and are thus no danger to the queen. This method is generally successful if the combs are placed next to the wall of the brood chamber. If the colony has brood in two or three hive bodies, the combs from the nucleus should be set in the top *brood* chamber next to the side. The bees of the colony gradually mix with those that were set in, usually without any fighting, and the queen continues to lay in her own combs, gradually extending her activities to other sections of the brood area in the hive. Separation of the transferred combs on the inner side, bottom, and top by a single thickness of newspaper may add to the success of this method, but if newspaper is used, examine the brood of the former queen a few days later for queen cells. The colony should be fed continuously.

Storage of Queens

There are times when queens are being produced faster than they can be shipped or otherwise used, and it is necessary to remove them from the nuclei and store them for a few days. Honey producers who receive queens from queen breeders also occasionally need to hold surplus queens for short periods before introducing them into their colonies.

The method of storing queens that is employed will depend on the length of time the queens must be held. Queens can be maintained in mailing cages, with attendants, for two weeks without apparent injury if they are kept a few degrees below brood-rearing temperature and have access to queen-cage candy *and water*. Water can be supplied with a small sponge on the screen, or with an inverted vial with perforated lid, or capillary tube that extends through the cage wire and into the chamber opposite the

Figure 76. Queens caged in two-hole cages without caged attendants. Cages in frames for nursery colony From *Contemporary Queen Rearing*, Dadant and Sons, Inc.

one with the candy.

A strong colony can control the temperature, the humidity, and the food supply for caged queens as well as, or better, than that which is provided in incubators. Consequently, surplus queens in cages without workers are stored for extended periods in strong *queenless* colonies. This can be done for a period of several weeks' duration without apparent injury to the queens if the "bank" colony is kept in good condition by regularly added emerging brood. The queenless storage colonies that are made up for that purpose are dequeened colonies, or are established with bees and brood taken from brood nests of good colonies. The cages of queens (Fig. 76) that are held in storage in colonies should not have attendants for the queen because the bees of the storage colony tend to be more antagonistic when attendants are included. The workers of the bank colonies will feed the queens through the wire screen of the cages and thus provide them with a more natural food than when they have access to queen-cage candy alone. The mesh of the cage screens must be no larger than 14 meshes per inch, or the cages must be double screened, to prevent bees from damaging the tarsi of the queens.

Special frames made to hold 48 three-hole or 78 two-hole queen cages placed back to back in tiers (Fig. 76) are suitable. Each queen reservoir or bank colony will care for three or four frames of queens. If queens must be held in storage during periods when the colony will not be strong enough to maintain a satisfactory cluster over all the queen cages, empty cages filling the ends of the rows of the holding frame will compact the distribution of the bees to cover the queens.

A few queens can be stored for a short time in an empty shallow body on top of a cell-building colony or on a strong colony in which the colony queen is confined to the lower story by an excluder. The candy end hole in each queen cage should be closed by a metal disc so the worker bees cannot release the queen by eating through the candy compartment.

CONTROLLED MATING

The female honey bee develops from a fertilized egg in which half of the hereditary factors is supplied by her mother through the egg nucleus and half by the father through the spermatozoon that unites with the egg nucleus. The male honey bee develops from an unfertilized egg and receives all of his hereditary factors from his mother, but she in turn received half of her factors from her mother and half from her father. It is apparent, therefore, that since half of the hereditary factors are passed on to succeeding generations through the male, the selection of the male parent is equally as important as the selection of the female parent if satisfactory progress in honey bee genetic studies and in developing honey bee strains for special purposes is to be made. Selection of the male parent means control of mating of the virgin queen.

Years of parental selection and control of reproduction in domestic animals and plants by humans have been major factors in the development of the numerous domestic animal races and plant varieties of the present time. Honey bees are an exception. Nesting in the wild and mating on the wing are inimical to selection and to breeding control. Beekeepers even now, for the most part, have lax control over the heredity of their bees.

The desire to control the mating of honey bee queens extends back at least to the 1700s and it continues to the present. Reaumur in 1740 confined a queen and some drones together in a glass dish expecting to witness the mating of the queen. This experiment failed, as have others of like nature since that time. Apparently the first person to try to inseminate a queen artificially was Huber in 1813, but he also failed in his attempt. In the years that followed, almost every conceivable way to inseminate the queen or to otherwise control her mating was tried. These attempts also failed, or at best, had indifferent success, and beekeepers returned their attention to ways to obtain desired matings with free-flying virgins and drones.

Isolated mating stations are presently being employed with

satisfactory results in some European countries (Dr. Jost Dustmann, 1994). Their usefulness in the United States is limited by the difficulty of finding suitable and sufficiently isolated areas, and by the expense of operating them. Extensive prairies, semi-deserts, elevations higher than feral bees can survive, latitudes beyond the normal habitat of honey bees, and islands, have been successful as isolated mating areas in North America, but their utility is relatively unexplored. The possibility exists that some of these can be used for commercial mass-mating of queens. As of today, it is necessary to saturate the air of the vicinity of the mating apiaries with the drones of the selected breeding stock to get reasonable mating control with *open mating*.

It appears that the surest way to control the male parentage of the breeding stock is to inseminate selected virgin queens artificially with spermatozoa from selected drones. The technique of artificial insemination (a phrase disliked by Dr. Lloyd R. Watson who substituted the phrase "instrumental insemination") has been developed during the past six decades to the point of practicality in routine breeding work. Yet, in spite of the advances made in techniques and instruments, and in reduction of costs, beekeepers have been slow to avail themselves of the benefits that instrumental insemination of queens can bring them.

Open mating of queens that will head honey producing colonies cannot be relegated to the past, however; it remains an integral part of modern beekeeping, but must be utilized with greater care in selecting and providing the proper mates for the queens, and in saturating the environment where mating takes place with the desired drones.

Notwithstanding its short history of success and its neglect by beekeepers, instrumental insemination of queen bees is firmly embedded in stock improvement programs and in stock maintenance. It is, furthermore, essential to many scientific studies that involve honey bees. Instrumental insemination and contemporary breeding methods give beekeepers control over the quality of their bees comparable to that employed by plant and animal breeders and, to the merely curious, a tool for genetic experimentation. For those beekeepers who plan to use instrumental insemination there are illustrated instruction manuals, photographic slide sets, and tapes available that present excellent instruction. These should be consulted. Some of the universities and colleges, and some of the bee culture laboratories of the United States Department of Agriculture occasionally conduct instrumental insemination instruction on an ad-hoc basis.

Possession of the power to control the matings of one's queens is gratifying. Nevertheless, it is recognized that many beekeepers would be dissuaded by the laboratory procedures that, though simple, require care and fine control of hand and finger movements, and by the demanding scheduling of drone and virgin rearing. But even those beekeepers should have interest in a biological procedure that, in insects, is limited to the honey bee.

A brief story of the development of instrumental insemination, and its present status, is presented in this chapter.

Instruments and Development of Instrumental Insemination

Until 1934, nearly all workers involved in the instrumental insemination of queen bees believed that by their method semen from the drone was injected into the reproductive system of the queen. It was the failure to recognize that semen was not being injected into the oviducts that resulted in continuing and sporadic partial success.

It was shown in 1911 (Zander) that the semen is deposited in the oviducts at natural mating; this fact was verified by other work in 1917 (Shafer), again in 1920 (Bishop), and in 1933 (Laidlaw 1944). When a queen mates naturally, the semen is deposited in the vagina, the median oviduct and the lateral oviducts, and the spermatozoa migrate to the spermatheca over a period of several hours as the queen pushes the semen backward into the sting chamber. The filled spermatheca contains 5,000,000, or more, spermatozoa, which live in the spermatheca for the life of the queen but do not increase in number there. A few spermatozoa are withdrawn in spermathecal fluid by the queen each time an egg that is to be fertilized is laid (Harbo 1979). After many eggs have been laid, the supply of spermatozoa may be so reduced that the queen becomes a partial drone layer. A queen with fewer than 5,000,000 spermatozoa in the spermatheca may be considered to be partially inseminated.

Mature drones can be made to evert the copulatory organ (*endophallus* or *penis*) and ejaculate the seminal fluids by pushing the abdominal segments together, by pressure on the abdomen, by decapitation, by pressure on the underside of the thorax, and by other stimuli. The most obvious method to inseminate the queen would be to bring the drone into juxtaposition with her and then cause the copulatory organ to evert into the sting chamber and vagina of the queen, followed by injection of the seminal fluids into the oviducts. When this was tried, the vagina received scarcely any semen. Most of the semen was forced into the bursa copulatrix and its pouches. This method was tried by many workers with only meager success: few sperm reached the spermatheca. The oviducts were not examined for semen because it was assumed that the semen had been deposited there.

Microsyringes and the Valvefold

R.L. Watson devised a microsyringe in 1926 that promised to afford an easier method than the foregoing for injecting the semen into the reproductive tract of the queen. Watson's syringe was constructed of glass, with a plunger extending into a capillary tip. The plunger was moved by a screw mechanism which gave positive control of its action. This syringe was efficient, except for the overlarge diameter of the syringe tip that prevented complete insertion of the tip into the vagina, and consequently semen leakage occurred and semen did not reach the oviducts. Syringes of various kinds had been employed unsuccessfully in efforts to inseminate queens before Watson, but Watson's syringe is the prototype of most of the syringes used after him.

Nolan (1932), using the Watson syringe as a model, constructed syringes from mechanical pencils and glass capillaries. These syringes were very good and inexpensive, but they, too, had the defect of a syringe tip that was too large to enter the vagina, and there was much leakage into the bursa copulatrix and its pouches, with little, if any, semen deposited in the oviducts.

Laidlaw showed in 1933 that a ventral inpocketing of the *vagina*, he called the *valvefold* (Fig. 77), prevented passage of the semen into the median and lateral oviducts, and that pushing this structure from over the opening of the median oviduct allowed semen to be injected into the oviducts (Laidlaw 1934). It was apparent that blockage of the median oviduct was even more a hindrance to instrumental insemination than semen leakage around the syringe. Laidlaw (1939) adopted Nolan's syringe, but tapered

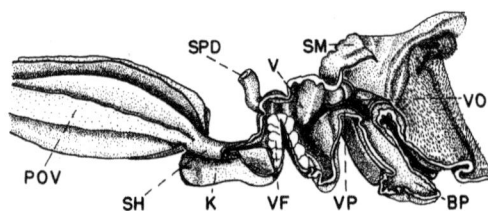

Figure 77. Drawing of dissection exposing valvefold. From Laidlaw 1939, 1944.

BP - pouch of bursa; K - keel of median oviduct; POV - paired oviduct; SH - shaft of median oviduct; SM - sting membrane; SPD - spermathecal duct; V - vagina; VF - valvefold; VO - vaginal orifice; VP - vaginal passage.

the end of the glass syringe tip so it could enter the vagina and seal it from leakage of semen. With this modification of the tip, semen leakage did not occur when the valvefold was pulled from over the median oviduct orifice. and semen was deposited in the oviducts, with the consequent migration of the spermatozoa to the spermatheca.

Mackensen (1948) designed and constructed syringes with removable glass or plastic tips which do not have the metal plunger in the tip. Instead, a liquid plunger is used, and the tip is tapered to enter the vagina. Mackensen's syringe is excellent and widely used, but is no longer available.

Schley (1982), designed and manufactures a finely crafted syringe with versatile capacity that is popular in Europe and is easy to use. Harbo (1985) modified an apparatus that was used by Poole and Taber (1969) to collect semen for a study of semen storage, to be an insemination syringe (Fig. 78). Harbo attached a fine glass syringe tip by a length of capillary tubing to a glass capillary, and connected the glass capillary with fine tubing to a Gilmont micrometer syringe. He enclosed the glass capillary in a larger glass tube that slides in the syringe holder of the Mackensen instrument for insertion of the tip into the vagina of the queen. This syringe has gained acceptance, and because of its simplicity and availability it may be one of the syringes of preference in the future. The Schley and Harbo syringes are equally suitable for insemination of virgin queens and are readily available.

Carbon Dioxide
It had been found in 1931 by Laidlaw that queens anesthetized with carbon dioxide or ether were easier to inseminate than those which were active during the operation (Laidlaw 1934). Mackensen (1947) demonstrated that queens subjected to carbon dioxide for two ten-minute periods are stimulated to lay whether inseminated or not. This discovery contributed a very important element to instrumental insemination. It is now known that two successive exposures as short as two or three minutes to carbon dioxide will cause the queen to lay. Ether has the same effect but the queens lay several eggs in a cell and miss cells.

Queen and Syringe Manipulators
The injection of the semen into the queen is a delicate procedure. The organs of the queen are so small (the vaginal orifice is about

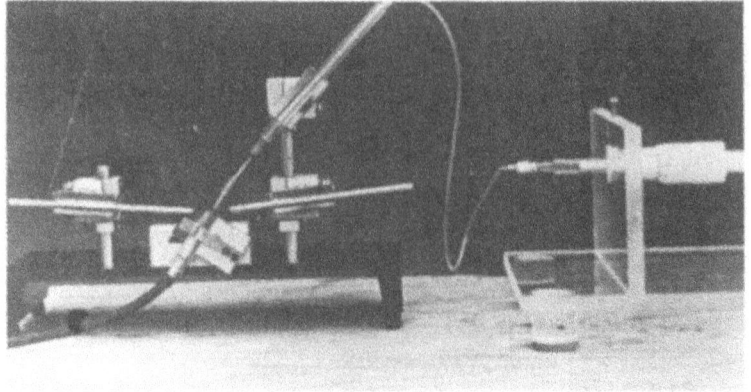

Figure 78. Harbo insemination set-up. Mackensen instrument, Harbo syringe and Gilmont micrometer. From Harbo, *American Bee Journal* 1985.

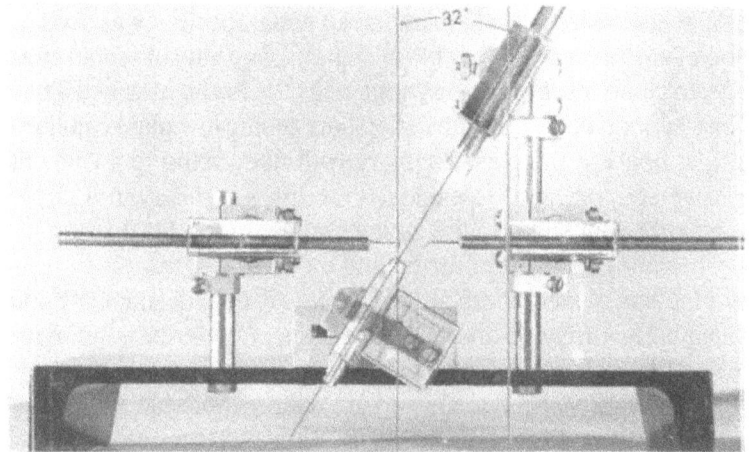

Figure 79. Mackensen insemination instrument. From Laidlaw, H.H. 1977. *Instrumental Insemination of Honey Bee Queens*, Dadant and Sons, Inc.

0.66mm. in diameter) that positive control of all movements is essential, and the queens must be immobilized. Watson tied the queen in a tilted "cradle" which he fastened to the microscope stage, and he clamped the syringe in a Barber pipette manipulator fastened to the right edge of the microscope stage. The manipulator held the syringe in line with the longitudinal axis of the queen and provided smooth positive movement of the syringe in all directions. The sting chamber of the queen was opened with forceps held in the operator's left hand and the syringe was inserted into the vagina by adjusting the manipulator with the right hand. Had not the valvefold interfered with passage of the semen into the oviducts, this technique and these instruments would have been quite successful when used by a skilled operator.

Chapter VI

Other workers developed their own apparatus and techniques. One of these, originated by Nolan and refined and improved by Mackensen and Roberts (1948), is the basic model for several others worldwide (Fig. 79). It consists of a heavy metal false stage which is made to lie transversely across, but unattached to, the stage of a dissecting microscope. It is fitted near both ends with metal posts which carry a metal bar between them to bear the queen confined in a plastic tube with her terminal abdominal segments protruding. The tube is provided with a longitudinally perforated stopper to which rubber tubing is attached for administering carbon dioxide to the queen. The tube is constricted enough at the upper end to prevent the queen from backing out of the tube.

Above the bar, each post is fitted with an adjustable metal block having a hole through which the handle of a modified teasing needle is inserted and in which it can slide. The modified needles, called ventral and sting hooks, are used to open the sting chamber. The sting hook is fashioned to fit beneath the sting and pull it from over the vaginal orifice. The syringe is held by an adjustable metal block secured to near the top of the post on the right side (for right-hand operation). Movement of the syringe for insertion into the vagina is accomplished by sliding the syringe through the hole in the movable supporting block. The valvefold is held from over the orifice of the median oviduct with a small thin probe while the end of the syringe is inserted into the vaginal orifice. Alternatively, the valvefold is pushed from over the median oviduct opening with the tip of the syringe; or the sting is pulled dorsally with the sting hook until the vaginal opening becomes a dorso-ventral slit; the syringe is adjusted to enter the vagina near the top of the slit and dorsal to the valvefold.

A simplified and inexpensive version (Fig. 80) of the Mackensen-Roberts instrument designed and made by Laidlaw and Goss (1990) can be made by beekeepers in their own shops. Most materials needed to make this device can be obtained from Small Parts, Inc, P. O. Box 4650, Miami Lakes, Florida 33014-0650. This instrument has the outline of the Mackensen instrument but differs in significant ways. The queen manipulator and the syringe manipulator are separate. The queen manipulator is movable on the microscope stage, or the supporting plate, and can be set aside while semen is taken into the syringe and then be re-

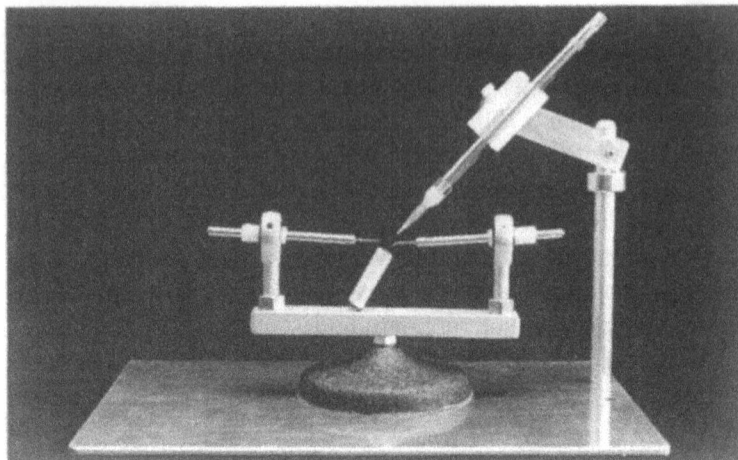

Figure 80. Laidlaw-Goss queen bee pre-set insemination instrument. Near side. Mackensen syringe. From Laidlaw and Goss, *American Bee Journal* 1990.

turned to the stage where it is aligned with the syringe by sliding it on the microscope stage, or on the plate.

The queen manipulator is made with an aluminum, or brass, bar 6"long, 3/4" wide, and 1/2" thick that is attached at its middle, by means of a 1/4" threaded rod, to the center of a heavy circular brass base that is 3 3/4" in diameter and 3/8" thick. The underside if the brass base is recessed to prevent rocking. The lower surface of the bar is 1 1/2" above the upper surface of the brass base to provide clearance for the operator's fingers between the bar and the instrument supporting base. At one end of the bar is a perpendicular 2.437" male end-rod that carries the rod-handle of the ventral hook, and at the other end of the bar is a like end-rod that carries the rod-handle of the sting hook. The queen is held in a plastic or glass tube that is fully open at both ends and is positioned so the terminal segments of the queen that protrude from the tube are centered over the middle of the bar. The queen holding tube is borne by a smaller metal tube that conducts CO_2 to the queen and is angled about 30° from the perpendicular. The queen is anesthetized before she is put into the queen holding tube and she is inserted head downward into the upper end.

The Mackensen and Roberts instrument is no longer commercially available but a wood and metal copy can be made by the beekeeper from directions given in *A Manual for the Artificial Insemination of Queen Bees* by Otto Mackensen and W. C. Roberts (1948), Publication ET- 250. USDA Division of Bee Culture. Directions for making Mackensen's all metal version is presented in

Instrumental Insemination of Queen Bees by Otto Mackensen and Kenneth W. Tucker (1970), Agriculture Handbook No. 390. ARS US Department of Agriculture, Washington, D. C. The Schley instrument, that is constructed on the Mackensen-Roberts model, is currently available commercially.

A significant departure from the generally preferred design of instrumental insemination queen manipulators is the Kühnert-Laidlaw instrument (Fig. 81) that was designed, and made especially for use with the Kühnert method of instrumental insemination. The sting hook is eliminated: the sting is pulled from over the vaginal orifice and maneuvered with forceps.

The base is a 60mm x 20mm brass disk. A blind hole is drilled 18mm in from one side of the disk toward the edge at an angle of 25° from perpendicular to receive a tubular stopper that supports the queen-holding tube. Another, directly opposite, is drilled toward the edge at an angle of 29°. In addition, two other blind holes, midway between the first two holes, are drilled toward opposing edges of the brass disk. One at 27° and the other at 32°. A side hole is drilled to intersect each stopper hole, and, when used, is given a fitting for attachment of Latex tubing for conduction of CO_2 to the queen. The queen is held in a section of plastic or glass tubing that slips onto the stopper, and a ventral hook is movably attached to the queen-holding tube. The queen, anesthetized or not, is put head downward into the tube. The sternum of the sting chamber is caught under the hook to hold the queen steady and the sting chamber open while the sting is held by forceps and the syringe tip is inserted into the vagina and the semen is injected into the oviducts.

The queen can be positioned under the syringe for syringe insertion by sliding the device, which is movable horizontally in any direction and can be rotated on the microscope stage. This instrument is the simplest of all of the instruments, the least expensive, and fully as efficient as any other.

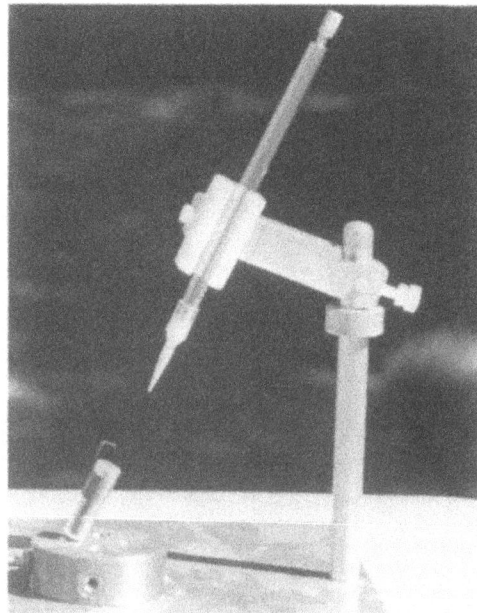

Figure 81. Kühnert-Laidlaw device for use with Kühnert method of instrumental insemination of queen bees. From Kühnert and Laidlaw 1994.

Drones for Insemination

Drones for insemination can be, and often are, in short supply when needed. Inseminations must be planned. Drone comb should be given to the drone mothers 35-40 days before the drones are needed. Drones are used when they ten days old. The drone mother colonies should be supplied with abundant pollen and be fed sugar syrup continuously as long as drones from them are needed. The drones must be emerged and matured in an otherwise drone-free colony or confined so they can be identified. The nursery, if drones are matured in one, is also generously provided with pollen and syrup. Newly emerged drones require the proper food to develop and mature the reproductive cells and organs, and nurse bees are needed for this.

Drones that are mature tend to congregate on the outer combs, and a partially empty comb at the side of the hive gives space for mature drones. Drones emerged free in a colony and marked at emergence may be permitted open flight. An empty hive body that is screened on top and is placed over their colony provides a cage for confined drones to have limited flight. Drones can be caged and matured in a nursery colony so the parentage of drones from many different mothers can be matured in a single colony.

Insemination Procedure

Instrumental insemination of queen bees can be learned from the instructions that have been published in detail. The discussion presented here is a brief overview of instrumental insemination to acquaint beekeepers and others with what instrumental insemination of queen bees actually is. Complete directions for using the Mackensen and Roberts apparatus are given in the two manuals mentioned above, and in Laidlaw (1976, 1977). The procedures in using the Mackensen-Roberts and the Laidlaw-Goss instruments are similar. Instructions for using the Kühnert-Laidlaw instrument are contained in Kühnert and Laidlaw (1994).

The following descriptions apply specifically to the Laidlaw-Goss instrument. To avoid overexposure of the queen to CO_2, the syringe is filled with semen *before* the queen is put into a queen-holding tube, or mounted in a queen holder of a queen manipulator. The Mackensen syringe is prepared for use by filling the syringe tip-connector-piece with a physiological saline solution containing an antibiotic, and the plastic syringe tip is screwed into the connector. By turning the syringe screw, some of the physiological solution is forced from the tip. The discharge is stopped when the fluid plunger will withdraw to the threads at the base of the tip when the screw is reversed. The fluid is again forced down the tip to within 3 mm of the end, and the syringe, now ready for use, is placed in its holder in the syringe manipulator.

Eversion of the copulatory organ is accompanied by ejaculation of the seminal fluids, and is brought about in a drone by mechanical stimulation. A drone is selected and decapitated. If this does not initiate eversion, the abdomen is drawn lightly over a towel or is squeezed with moderate pressure (*do not mash*) between the thumb and forefinger, or by pushing the abdominal segments together. If the abdomen contracts under one of these stimuli, ejaculation has probably occurred, but eversion usually stops before completion and before the semen has been released to the exterior of the endophallus. Moderate pressure on the drone's abdomen and base of the partially everted penis will continue eversion and release the semen and accompanying mucus to the exterior (Fig. 82). Too much pressure, however, may cause the penis to rupture, with consequent lost of the semen.

The drone, with the everted organ and adhering fluids, is brought to the tip of the syringe. With the semen *just touching* the syringe tip the semen is taken into the tip. Care must be taken to avoid touching the white mucus; it will clog the syringe. If more than one drone is used, this procedure is repeated until the desired amount of semen has been taken into the syringe.

Absolutely no fecal matter from the drone, or queen, or hands should be taken into the syringe, and the semen should not be permitted to touch any part of the drone's body. *Any contamination may result in the death of the queen* If the tip needs to be cleaned on the outside, this should be done with clean, sterile tissue, never with bare fingers. It is good practice to wash the hands frequently, and always when the tips are changed.

Queens, drones, and workers accumulate feces when caged. A queen taken from a nursery cage for insemination, and placed in a deep wide-mouth jar will void the feces as she crawls and flies up the sides of the jar. Drones taken from nursery cages will void feces if allowed to crawl and fly in a drone flight-box.

Eight cubic millimeters of semen are taken into the syringe for most inseminations, and semen from at least eight to twelve drones is required. After the semen is taken from a drone, it is withdrawn one millimeter into the syringe tip to prevent drying before taking semen from the next drone. Some of the translucent mucus or watery bulb fluid may be taken up with the semen

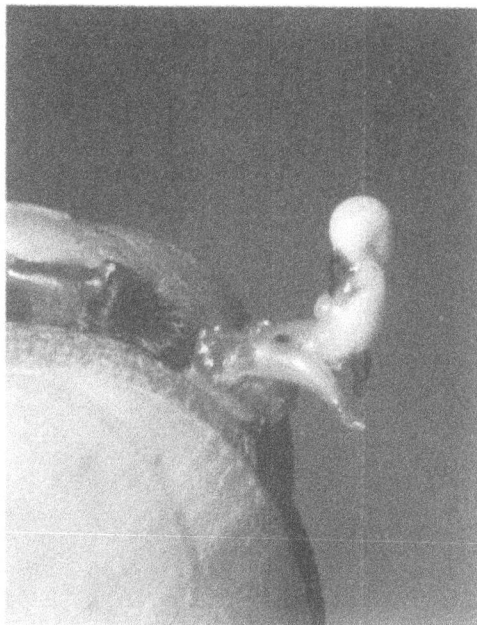

Figure 82. Semen at end of everted penis covers mucus from the mucus glands. From Laidlaw 1977. *Instrumental Insemination of Honey Bee Queens*, Dadant and Sons, Inc.

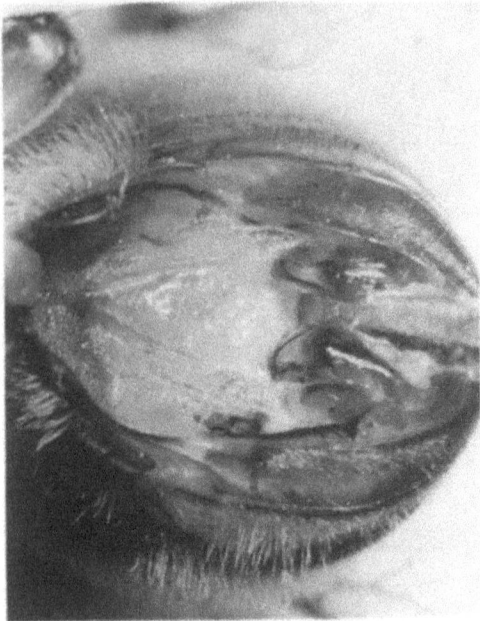

Figure 83. Opening vaginal orifice. Sting hook properly placed against sting. <-shape ventral edge of bursa copulatrix visible in center of opened sting chamber. From Laidlaw 1977.

to prevent drying of the semen at the end of the syringe, but none of the white mucus should be included because it will coagulate in the syringe and plug the median oviduct. After the syringe is filled with semen, about a millimeter of the fluid plunger is taken into the tip to be released to lubricate the tissues as the syringe is inserted into the queen reproductive tract. The syringe is then raised as far as the syringe manipulator will take it, so that it will clear the queen manipulator hooks.

The queen is anesthetized to motionlessness, a matter of about one minute, in a small separate container into which CO_2 is flowing slowly from a rubber tube inserted into the side of the container near the bottom. She is then placed head downward into the queen holding tube, and carbon dioxide is adjusted to flow slowly into the queen-holding tube. The opening hooks are brought into position: first the ventral hook to hold the queen's abdomen firmly, and then the sting hook. The sting hook is fitted in the triangular area between the bases of the sting lancets, and the sting is pulled dorsally from over the vaginal orifice (Fig. 83).

The syringe is lowered into the sting chamber and the queen manipulator is moved on the microscope stage until the end of the syringe is over the vaginal opening (Fig. 84). The syringe is lowered to just enter into the vaginal orifice, and the queen manipulator is moved slightly to the left and then slightly to the right

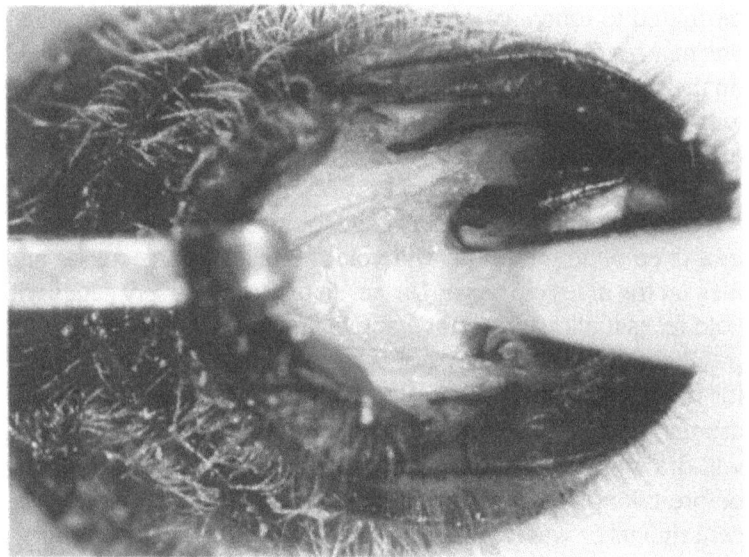

Figure 84. Inserting syring tip into queen reproductive tract. From Laidlaw 1977.

by sliding the queen manipulator on the microscope stage, and the syringe is inserted into the reproductive tract about 1mm (Fig. 84) or until the membrane surrounding the syringe moves slightly with the movement of the syringe. The correctness of the position of the syringe is tested by a slight turning of the syringe screw. If semen moves down the syringe tip easily, then the tip is placed correctly. If not, the syringe is improperly placed, and the syringe is readjusted, or semen has dried in the tip and the syringe must be withdrawn and cleaned. The semen is injected, including the air bubble that separated the semen and the liquid plunger (Fig. 85), and the syringe is raised and the fluid plunger run to the end of the tip to prevent semen from drying on the inner wall of the tip.

It is, perhaps, somewhat easier and faster to open the vaginal orifice by manipulating the sting with the sting hook until the normally transverse orifice becomes a ventro-dorsal slit. The syringe is then inserted into the orifice about one millimeter or until the membranous tissue around the vaginal orifice begins to move. If the sting is pulled upward and dorsally with the sting hook the vaginal orifice moves with it. The syringe can be slipped into the vagina at a slant.

The hooks are now disengaged and moved up and away from the queen. The queen manipulator is again set aside and the queen is removed from the queen holder. The *tip* of one wing is clipped to indicate she is instrumentally inseminated.

Syringe tips are sterilized after use by injecting 6% sodium hypochlorite solution into them with a medicine dropper and thoroughly flushing them with distilled water. It is imperative to keep the equipment clean. The valvefold probe is polished frequently with jewelers' rouge, and no wax or other substance of any kind is permitted on the stage; a minute fleck of wax will interfere with the smooth movement of the queen manipulator over the stage.

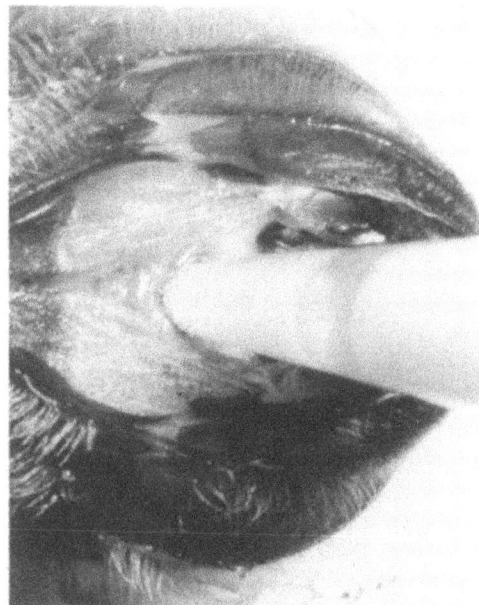

Figure 85. Injection of semen nearly completed. From Laidlaw 1977.

Results

Queens that mate naturally receive semen from many drones. Usually, naturally mated queens receive a normal insemination. Mackensen and Roberts determined that the spermathecae of naturally mated queens contain an average of 5.73 million spermatozoa, ranging from 3.34 million to 7.35 million. Mature drones were found to have an average of about ten million spermatozoa in the seminal vesicles, more than enough, therefore, to fully inseminate one queen. Their studies revealed further that the spermathecae of queens inseminated artificially with semen from one drone contained an average of 0.87 million spermatozoa, while the spermathecae of queens inseminated once with the semen from several drones had 2.5 cubic millimeters of semen; (the best drones had nearly 1 cubic millimeter of semen) containing an average of 2.97 million spermatozoa, and queens inseminated twice with semen from several drones received an average of 4.11 million spermatozoa in the spermatheca.

It is apparent from the above data that queens inseminated by present methods must receive semen from several drones if inseminations approaching the normal are to be obtained. Queens which are to maintain full colonies must have at least nearly normal inseminations. Queens inseminated at the University of California, Davis, were for many years inseminated once with at least 5 cubic millimeters of semen, and it has been routine for several decades to inject about 8 cubic millimeters. These queens have headed field colonies as successfully as naturally mated queens. Naturally mated queens, as a rule, begin to lay about three days after mating; a few starting at two days and some at four days or longer. Artificially inseminated queens, not treated twice with carbon dioxide, sometimes do not begin laying until a week or more after they are inseminated. Queens inseminated at Davis in the 1950's sometimes delayed oviposition for an abnormal period, but a good proportion also laid fairly promptly. In one lot of 132 queens inseminated once in 1949 with 2.5 cubic mm of semen, and with one CO_2 exposure which was administered during insemination, 9 died of paralysis or other causes, 17 were given a second CO_2 treatment 3 to 17 days following insemination after which they laid within two to five days, and 106 laid without a second insemination or CO_2 treatment. Of these, five (4.1 per cent of the 122 queens which laid) began oviposition the second day following the insemination, 29 (23. 8 per cent) laid at three days, and a total of 84 (68.5 per cent) were laying by the seventh day.

The discovery that carbon dioxide will cause the queen to lay has, to a large extent, solved the problem of delayed oviposition; nearly all queens given two treatments of CO_2 will lay within ten days, and very often much sooner. Why some inseminated queens lay promptly without a second CO_2 treatment and others do not, remains to be discovered. Whether an inseminated queen will initially produce all worker brood or a mixture of worker and drone brood in worker cells appears to have little relation to the number of spermatozoa in the spermatheca, except that queens which have normal or nearly normal inseminations almost invariably produce all worker brood. Many queens that received few sperm in the spermatheca nevertheless produced all worker bees. In a few other instances, with the spermatheca fairly well supplied with spermatozoa the queens were partial drone layers. The percentage of worker brood is thus an unreliable measure of the relative numbers of spermatozoa a queen possesses. Queens inseminated with 5 cubic millimeters or more of semen will usually be capable of maintaining a large colony for a season. Queens with considerably less than a normal insemination may become partial drone layers after a heavy and prolonged period of brood rearing.

Little is known of the physiological factors involved in the insemination of queen bees. There is evidence which indicates that drones must be well fed and cared for the first few days after their emergence if they are to mature large numbers of sperm. Drones may possibly differ not only in the quantity of semen, but also in the viability and activity of the spermatozoa. Drones matured in cages in cell-building colonies where they receive good care are ready for use when they are ten days old. Queens may be emerged either caged or free in the nuclei they are to head; queens emerged in nursery colonies, inseminated and then introduced into regular colonies perform as well. Emerging queens in cages in nursery colonies during inclement weather has a distinct managerial advantage over emerging them in nuclei, but there are also some disadvantages such as loss of inseminated queens in the nurseries or damage to them, and introduction losses.

Care and Introduction of Inseminated Queens

A virgin queen emerged free among the bees of her colony is already introduced, and a virgin queen emerged in a cage in her colony is also introduced. Whenever possible, or practical, virgin queens to be inseminated should be emerged caged in their nuclei and marked or clipped at the time of insemination. There is then no doubt of parentage. If virgins cannot be emerged in their own colonies, queenless nursery colonies are used to emerge cells and care for virgin queens. Nursery colonies are also used to mature and care for drones, and to care for inseminated queens.

Nursery colonies are one or more stories high. The bodies contain nine frames so arranged that there is a frame of honey and pollen next to each sidewall and one in the center. A frame of larvae occupies the third position from each side. This arrangement provides for four frames of caged queens or drones, one on each side of the combs of larvae. Package bees are added to make the colony strong, and new larvae and package bees are added each week. The colonies are fed continuously with sugar syrup, or by means of candy in a feeder screen placed over the top of the hive like an inner cover.

Virgin queens and drones need no particular attention in the nursery colony, except that the bees should have access to the drones through an excluder on one side of the cage. Inseminated queens present a problem. The bees often show antagonism toward them, and unless the cages are properly constructed with 14-mesh screen, or the cages are double screened, the queens may lose tarsi or segments of the antennae. It is not unusual for the bees to neglect some of the queens so that they starve. This can be avoided by putting candy in the cage with the queen when she is returned to the nursery.

Queens inseminated from nurseries are left in the nursery at least over night. They are introduced into newly established regular nuclei, or into five-frame nuclei in full bodies which have been made up in such a way that the older bees return to the parent colonies. A Boardman entrance feeder is placed on the bottom board of a five frame nucleus in the space where the remaining frames will go later and it is provided with a quart jar of light sugar syrup. Feeding is continuous until the queen begins to lay.

The queen is put into the nucleus in a Miller introducing cage or some modification of it, such as the JZ^s BZ^s cage, or under a push-in cage, or in a California mini queen cage. Several attendant bees from the *receiving* nucleus that are *swollen with honey or syrup* can be put in the cage with the queen without danger to the queen. The entrance to the nucleus is covered with queen excluder until the queen lays. The excluder is then removed from the nucleus and is replaced with a robber screen.

It should be remembered that scarcity of old bees, and a constant supply of clean, fresh syrup, or at least a light honey flow, are essential to safe introduction of inseminated queens.

GENETICS Chapter VII

Superior queens cannot be produced by good queen-rearing methods alone. The quality of the stock is equally as important as the manipulative practices. This chapter and Chapter VIII are concerned with stock improvement. To the beekeeper who has had no formal biological training these chapters may be somewhat difficult to understand, for they involve terms and material with which the beekeeper is ordinarily not familiar. They are not, however, beyond the grasp of anyone who will study them carefully, and we believe that the beekeeper who is sincerely interested in improving his stock would do well to make a serious study of these chapters on breeding and stock improvement.

Technical terms used in these discussions will give the reader little difficulty if he or she bears in mind that technical words are nothing more than names, and that they are often the only names that are suitable. If technical words were not used, it would be necessary to repeat tedious descriptions that would complicate the narrative and make for monotonous reading. If the meaning of a technical word is learned when the word is first encountered, the term quickly becomes a part of the vocabulary of the reader, and as vocabulary grows reading and understanding technical descriptions and discussions become less difficult.

Breeding, Selection, and Inheritance

Breeding, as used in this book, is the systematic mating of selected individuals to produce offspring that possess characteristics considered desirable. The matings are made, therefore, with a definite goal in view, and purposeful matings are continued through successive generations. Breeding differs from *rearing* very clearly: rearing is the production of individuals from the egg to the adult, and is not necessarily concerned with parentage of the individuals produced, nor with their characteristics or the characteristics of their offspring.

Selection involves evaluation of individuals based on the determination of their characteristics and of their ability to transmit

Genetics   141

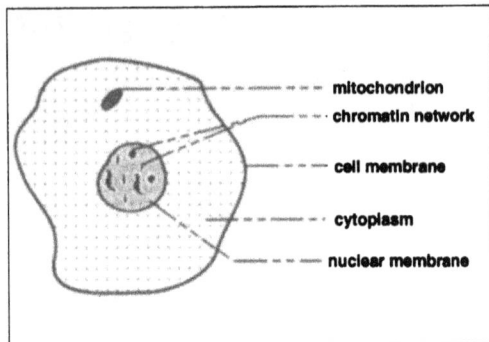

Figure 86. A cell.

factors that produce these and other characteristics in their offspring. Selection is an important part of any breeding program. The transmission of factors from parent to offspring is called *inheritance*.

The sequential procedures of selecting, breeding, and again selecting is much the same as those for other organisms. All are subject to the same basic laws and mechanisms of heredity that govern passage of various features from parent to offspring. It is desirable to examine these mechanisms before breeding methods can be understood and intelligently applied.

The Physical Basis of Inheritance

All organisms are made up of *cells*, small units of living material, and their products. The bee's body is composed, in part, of a very great number of cells; cell-manufactured substances constitute most of the rest of the body. The body wall of a bee, for instance, was manufactured by the layer of cells which lie against the inner surface.

Although different cells of a bee have taken on various forms and functions, each is composed of a basic viscous complex mixture of substance called *protoplasm*, which is encased in a membrane. The protoplasm is divided into *cytoplasm* and a more-or-less spherical *nucleus* within the cytoplasm (Fig. 86). The nucleus is likewise encased in a membrane. The cytoplasm has various bodies and inclusions, which we need not consider in detail here. Our main concern is with the nucleus.

The nucleus contains a material which stains more deeply with certain dyes than do other cell parts. This material is arranged in a network throughout the nucleus. It is called *chromatin*, and it bears the *genes* (factors), too small to be seen with a microscope, which in cooperation with the environment determine which characters will appear in the offspring.

The genetic information that orchestrates the development of an individual bee from a zygote formed from the fusion of an egg and a sperm cell, into an adult, is contained in the chromatin of the nucleus. Heritable differences among individuals are a consequence of having differences in their genetic codes. The genetic code is a sequence of submolecules, called *nucleotides*, that are arranged in sequential order into a much larger molecule called *deoxyribonucleic acid* (DNA). There are four kinds of nucleotides in DNA: thymine (T), adenine (A), cytosine (C), and guanine (G). The nucleus of a honey bee cell contains deoxyribonucleic acid (DNA) molecules consisting, in total, of about 180 million sequential nucleotides. Embedded in this very long sequence are the coded instructions for the production of *proteins* that are responsible for building a honey bee. Instructions are packaged in sets of three nucleotides that make up the genetic code and specify specific *amino acids* that are the building blocks of proteins. For instance a sequence of CCG-CTG-CAC-TGA would code for the amino acid sequence glycine-aspartic acid-valine-threonine, in a protein.

Another membrane bound intracellular body is the *mitochondrion*. Each cell contains several mitochondria that act as the power plants of the cells. Within each mitochondrion is a circular piece of DNA that contains the genetic codes for genes that are specific and necessary for the functioning of the mitochondria. Although these genes are not directly involved in determining characters of importance to bee breeding, they are of interest to breeding because they are inherited only from the mother and can be used as indicators of the maternal origins of stocks, for instance determining African from European bees. However, our main concern for bee breeding is with the genes that are contained within the nucleus.

The growth of an individual involves an increase in the number of cells of the body. Each cell must be complete and must have the same genes as every other cell. Thus, honey bee growth and development requires some mechanism which produces an increased number of cells, each provided with an identical chromatin network. This is accomplished by each cell dividing into two, a process called mitosis. In turn, each daughter cell divides, and this process continues until the pattern of development is complete.

When a cell divides by mitosis (Fig. 87), the strands of the chromatin network thicken, forming a long thread which finally separates into individual segments known as *chromosomes*. Sometimes the thread splits longitudinally before the chromosomes become recognizable; if not, the chromosomes split longitudinally after separation. In either event each chromosome has duplicated itself so the cell now has twice the number of chromosomes as before. The membrane of the nucleus disappears and structures that look like contractile fibers form a spindle-shaped pattern

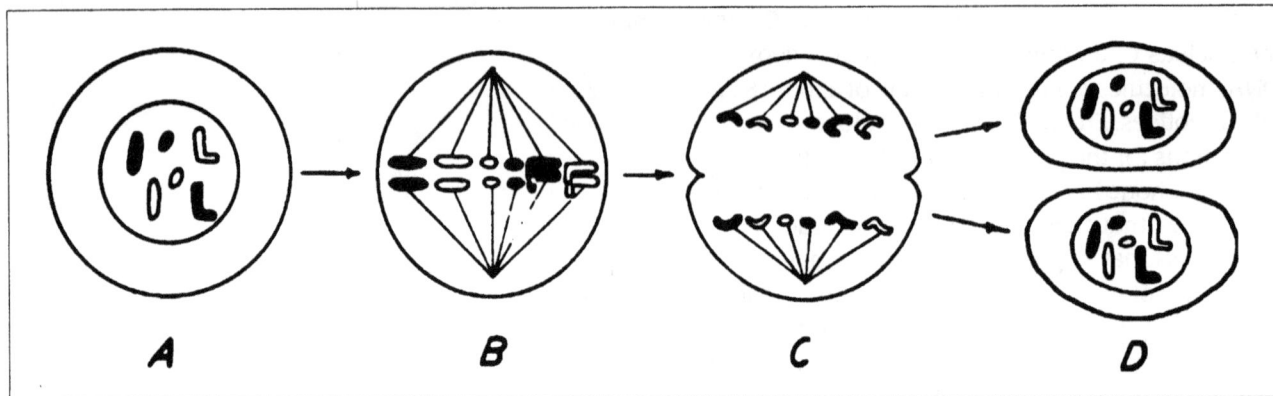

Figure 87. Mitotic cell division (diagrammatic). A. Cell with the full number of chromosomes (represented in this diagram by six, or three pairs). The solid chromosomes may be considered to have come from the queen's mother and the lighter chromosomes to have come from the queen's father. Thus each chromosome pair is made up of a chromosome from the queen's mother and the queen's father. B. The nuclear membrane has disappeared, a spindle has formed, the chromosomes have arranged themselves across the spindle, and each has split lengthwise. C. The halves of each chromosome have separated and have gone to opposite ends of the spindle. D. A nuclear membrane has formed around each group of chromosomes and the cell has divided into two cells. Each daughter cell has exactly the same kind of chromosomes.

across the nucleus, with the ends lying in the cytoplasm of the cell. The chromosomes arrange themselves across the middle of this spindle and the halves of each doubled chromosome separate and go to opposite ends of the spindle. A membrane appears around each group of chromosomes, forming a nucleus, and a new cell membrane, which develops across the middle of the spindle, separates the two new nuclei into new cells. In each, the spindle disappears and the chromosomes go through the reverse process to form a thread and a chromatin network. This process has the utmost significance. By it a single cell divides into two cells that have exactly the same hereditary factors because each daughter cell gets half of each chromosome; chromosomes carry the factors—genes.

The cells continue to divide as the organism develops, and the resulting cells come to perform different duties. In the bee, some form the brain, others the intestine, and still others form the layer of cells which produce the exoskeleton, or body covering. Certain cells, however, are set aside early in the development of the bee; they take no part in the development of the bee itself but are reserved for the propagation of the next generation. These reproductive cells are called *germ cells*. They eventually come to lie in the *ovaries* or *testes* of the bee.

The Germ Cells

The germ cells of the queen honey bee develop into the eggs, and the germ cells of the drone develop into the *spermatozoa*, known collectively as *gametes*. The queen may lay many thousand eggs during her lifetime; the drone produces about ten million spermatozoa. The mature gametes are derived from germ cells located in the closed ends of the ovarioles of the queen's ovaries and the closed ends of the testes tubules of the drone. The germ cells increase in number by mitotic cell division, but before they become gametes and are functional they must undergo a process called *maturation*.

The body cells and the unmatured germ cells of a queen (or worker) honey bee have 32 chromosomes which, in the non-dividing cell, are arranged into the chromatic network of the nucleus. These chromosomes are *paired*; that is, each chromosome has a mate of identical size and shape. One member of the pair came from the mother of the female bee, through the egg, and the other member came from the father, through the spermatozoon which fertilized the egg. The two mates of a chromosome pair are called *homologous* chromosomes; each chromosome is the *homologue* of the other.

There are then 16 pairs of homologous chromosomes in the cells of the female bee. These form two *sets* of 16 chromosomes each, a set consisting of one chromosome from each pair. Such an organism is said to be *diploid*. (The drone, which has one set, is said to be *haploid*.) When the egg matures, which occurs about the time it is laid, one member of each pair of chromosomes is eliminated from the nucleus, leaving the egg with one chromosome set. This is accomplished by a process of division of the nucleus of the egg called *meiosis*.

Meiosis in the bee egg consists of two successive nuclear divisions (Fig. 88), which occur in a sequence similar to that of mitosis. Chromosomes develop from the chromatin network, the nuclear membrane disappears, and division spindles form. However, meiosis takes place in the egg about the time it is laid, and the egg itself does not divide into two eggs. There are, therefore, nuclear divisions without complete cell divisions. Moreover, the chromosomes of maternal origin and the chromosomes of paternal origin at this time come close together on the spindle of the first division, forming 16 *pairs* of homologous chromosomes. This intimate pairing of homologous chromosomes is known as *synapsis*. During synapsis, or prior to it, the homologous chromosomes of each pair split, yielding four chromosomes, instead of two, closely associated as a group. Each such group is a *tetrad*. At

Figure 88. Meiosis in the bee egg (diagrammatic). A. Nucleus of egg with full numbers of chromosomes. Chromosomes represented by six, or three pairs. B. Chromosomes of each pair have come together on its spindle and each has split thus forming tetrads. C. The four chromosomes of each tetrad separate by pairs, two chromosomes of each tetrad going to one end of the spindle and two going to the opposite end. D. Two groups of chromosomes result. E. A spindle forms across each group of chromosomes and the two members of each chromosome pair separate and go to opposite ends of the spindle. F. Four nuclei are formed. One is the pronucleus of the egg and survives. The other three are polar bodies and disintegrate.

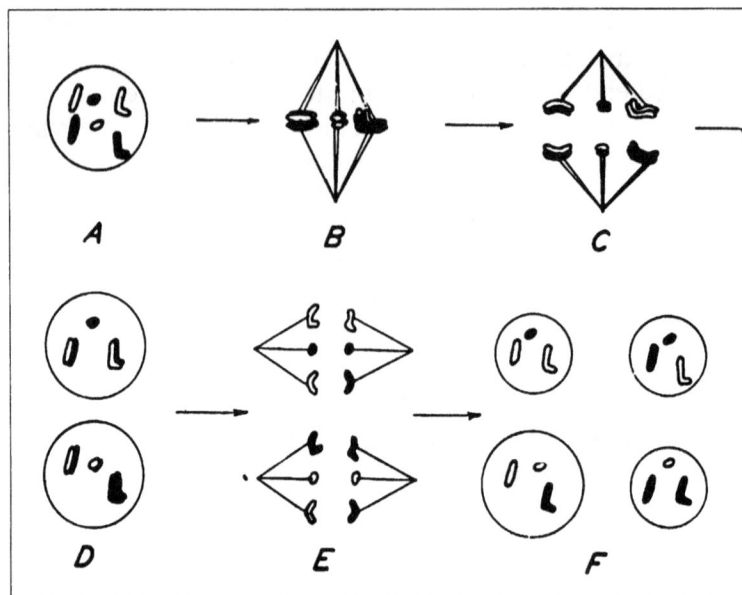

this time there may be an exchange of segments between homologous chromosomes, chromosomes from the mother exchanging segments with the chromosomes from the father. This exchange is known as *crossing over*, or *recombination*, the mechanism whereby groups of genes on a particular chromosome can be exchanged for groups of similar genes on its homologue. Pairing, breaking, and rejoining of chromosomes, the separation of the two members of each pair of homologous chromosomes during meiosis (Fig. 89), and the random assortment of chromosomes into daughter nuclei (Fig. 90), make possible many different genetic combinations in the pronuclei of eggs. It is highly improbable that a queen will ever produce 2 eggs that contain the same genetic information.

Two of the chromosomes of each tetrad now go to each end of the division spindle (Fig. 88). A new spindle appears across each chromosome group, and the two chromosomes of each tetrad which had gone together to one end of the first division spindle now separate and go to opposite ends of the new spindle. This results in four nuclei in the egg, all of which have half the original number of chromosomes, but with a chromosome from each original pair. Only one of the four nuclei survives; it is called the egg *pronucleus*. The other three, known as *polar bodies*, disintegrate.

Chapter VII

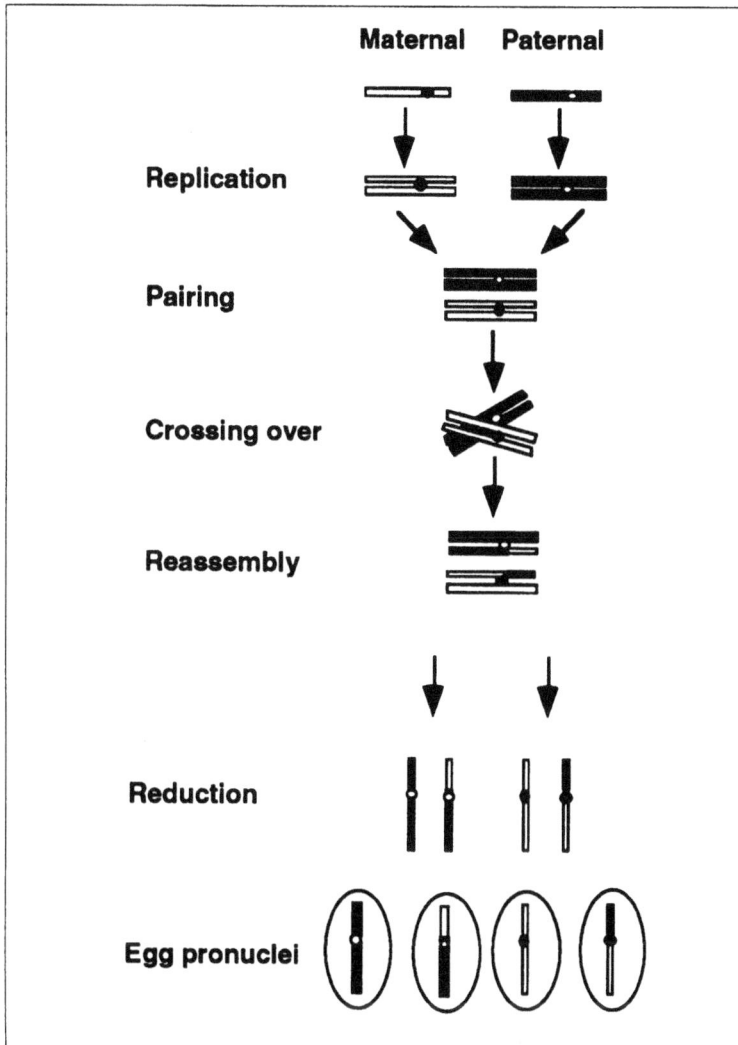

Figure 89. Genetic recombination during meiosis in the queen (only one of the 16 pairs of chromosomes is represented here). Primary cells that eventually develop into eggs begin with two sets of chromosomes; one inherited from the mother (represented by the open bar) and the other from the father (represented by the black bar). During meiosis, chromosomes replicate themselves, pair, break, and rejoin, then individual, single sets of chromosomes are assorted into separate egg pronuclei.

The egg pronucleus unites with a sperm nucleus if the egg is fertilized, or develops independently if the egg is not fertilized. In either case it gives rise to the bee larva by repeated mitotic divisions.

As mentioned above, the 32 chromosomes of a female bee are paired. One member of each pair comes from the mother of the queen through the egg, the other member from the father through the spermatozoon which united with the egg. At meiosis it is a matter of chance which mate of a particular chromosome

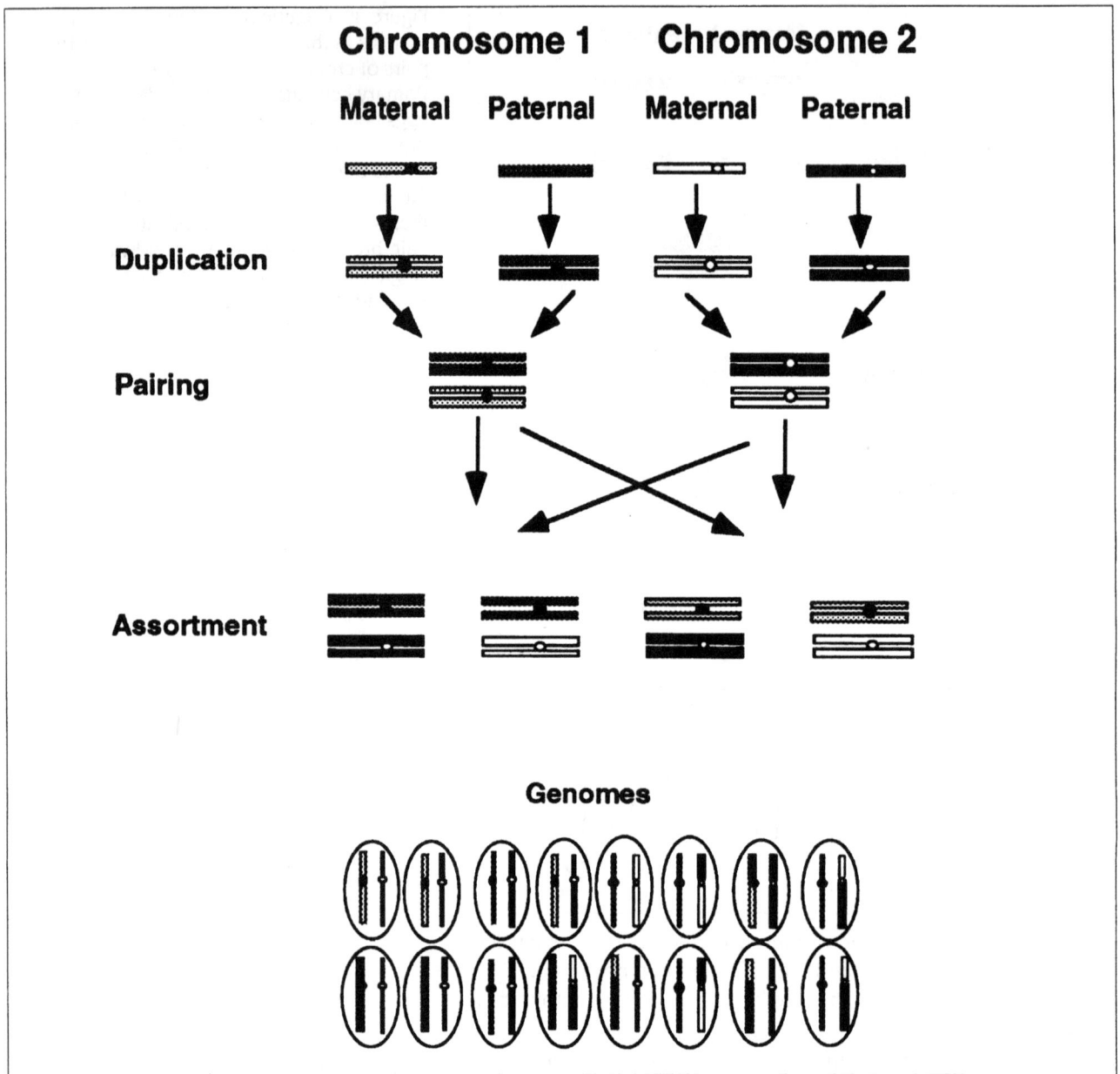

Figure 90. Independent assortment during meiosis in the queen (two of 16 pairs of chromosomes are shown). Homologous chromosomes derived from the different parents assort into egg pronuclei independently of each other. In combination with recombination shown in Figure 89, an enormous number of different genomes are possible, making every individual produced one of a kind.

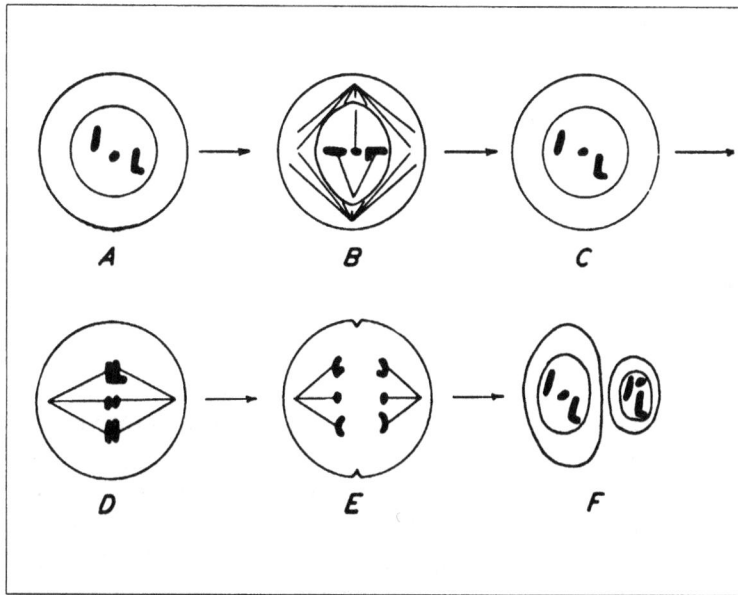

Figure 91. Maturation division during formation of spermatozoa in the bee (diagrammatic). A. Early reproductive cell of drone. Drone primary sex cells have half as many chromosome as cells of the queen or worker and in this diagram only three chromosomes are represented. B. Division spindle forms but nuclear membrane does not disappear, and C. cell does not divide. D. A second spindle forms, the chromosomes split and arrange themselves across the spindle, and the nuclear membrane disappears. E. The halves of each chromosome separate and go to opposite ends of the spindle. F. Two cells are formed each with the same number of chromosomes as the original cell. One cell is much larger but both eventually develop into spermatozoa.

pair goes to one or the other end of the division spindle with a particular mate of another pair, but the two mates of any one pair never go to the same end of the spindle. They always separate and go to opposite ends.

The drone develops from an unfertilized egg which has only one set of 16 chromosomes. As a consequence, germ cells of the drone have 16 chromosomes. Before male germ cells can fertilize an egg, they must undergo development from the early male germ cells, which resemble morphologically unspecialized cells, to spermatozoa, each of which has a narrow *head* containing the chromosomes in a greatly condensed state and a long thin *tail* which by a waving motion propels the spermatozoon. During the course of this development, incomplete meiotic divisions occur (Fig. 91). A spindle forms and the chromosomes arrange themselves across the spindle and move apart, but the nuclear membrane does not disappear and the nucleus does not divide. Thus the nucleus keeps its full set of 16 chromosomes. Immediately following this abortive reduction division, another spindle forms at an angle to the preceding one. The chromosomes split and arrange themselves across the spindle, the nuclear membrane disappears, the chromosome halves separate and move toward opposite spindle ends, and the cell divides into two cells with 16 chromosomes in each.

One of the cells initially retains most of the protoplasm of the parent cell, however, both cells develop into spermatozoa.

Fertilization of the Egg

The uniting of a male and female gamete is called *fertilization*. The resulting cell is a *zygote*. The upper end of the bee egg has a minute opening, the *micropyle*, that serves as an entry way for the spermatozoa. Spermatozoa are stored in the *spermatheca* of the mated queen. A few spermatozoa at a time are withdrawn from the spermatheca and sent down the spermathecal duct by the action of the sperm pump. When the egg is laid by the queen, it passes the end of the spermathecal duct on the way through the vagina. If it is destined to become a female bee, it apparently is held in the vagina for an instant while the micropyle is pressed against the opening of the spermathecal duct. One or more spermatozoa then enter the egg and the egg is deposited in the cell.

After the spermatozoa enter the egg they undergo a transformation (Fig. 92). The tails disappear along with the mitochondria of the sperm, the heads enlarge, and they again take on the appearance of nuclei. They spread out in the upper end of the egg and move downward in the cytoplasm while the egg is undergoing maturation. As soon as the egg maturation divisions are finished, the pronucleus begins to move across the egg cytoplasm and it meets one of the sperm nuclei in its path. The two nuclei

Figure 92. Fertilization of the bee egg (diagrammatic). A. Egg, prior to being laid, with full number of chromosomes (only 6 chromosomes or 3 pairs are shown). B. Maturation divisions in progress as egg is laid. Spermatazoa enters egg. C. Maturation divisions finished. Pronucleus now has only half the original number of chromosomes. Sperm head has again taken on appearance of nucleus. D. Sperm nucleus and egg pronucleus meet. E. Chromosomes of the two nuclei split and arrange themselves across spindle. Halves of each chromosome separate and go to opposite ends of the spindle. F. A nuclear membrane has formed around each group of chromosomes and there are now two nuclei each with the full number of chromosomes.

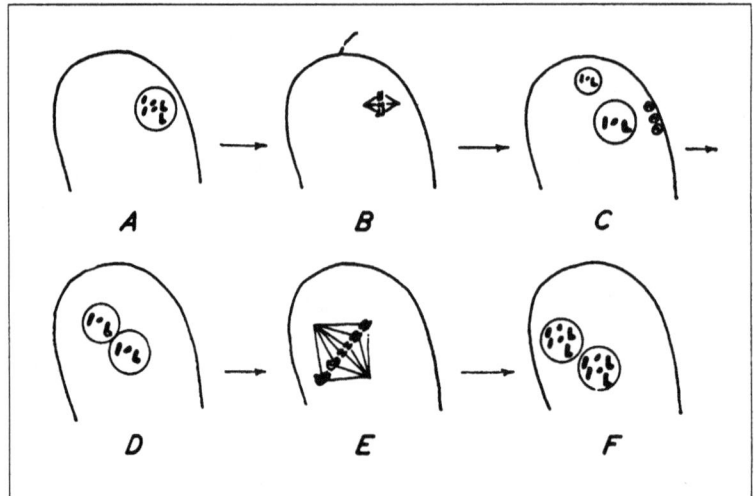

Chapter VII

come close together, a division spindle forms across them, the nuclear membranes disappear, and the chromatin networks give rise to replicated chromosomes that separate and align themselves on the spindle. The separated chromosomes then go to opposite ends of the spindle to form two nuclei with identical chromosome sets, and the development of the bee has begun. The reduction in the number of chromosomes from 32 to 16 during the maturation of the egg furnishes the mechanism whereby the fertilized eggs always have 32 chromosomes. The male and female gametes each had 16 chromosomes; their union restored the chromosome number in the zygote to 32.

If the egg is destined to become a male bee, it does not receive spermatozoa as it passes through the vagina. The pronucleus of the matured egg reaches the opposite side of the egg without encountering a sperm nucleus. It then divides by mitosis to begin the development of the drone. The matured egg nucleus had a single set of 16 chromosomes, and, since none have been added, the drone also has 16 different chromosomes. However, most of his tissues have more than 16 chromosomes due to replication of the original set into multiple copies, a process called *endomitosis*. The development of an egg without fertilization is known as *parthenogenesis*.

Parthenogenesis

Parthenogenesis, the origin of an individual from an unfertilized egg, is the hallmark of the insect order Hymenoptera that is composed of the bees, wasps and ants. Males are produced from unfertilized eggs and, except in special circumstances, have no father, a kind of parthenogenesis called *arrhenotoky*. Occasionally, females (workers and queens) are produced from unfertilized eggs, a process called *thelytoky*. A combination of the processes that lead to arrhenotoky and thelytoky sometimes result in strange genetic aberrations of workers and drones in hives.

Arrhenotoky

In 1845, Johann Dzierzon published his theory that male honey bees are derived from unfertilized eggs. He proposed that a drone has a mother, but no father. In 1913, H. Nachtsheim demonstrated that female-destined fertilized eggs have 32 chromosomes (16 pairs) while drones have just 16 chromosomes composed of one member of each chromosome pair. Subsequently, it has been shown that this form of parthenogenesis is characteristic of the Hymenoptera. Males arise from unfertilized eggs laid by either queens or laying workers while females are normally biparental. Eggs of honey bees, therefore, have a peculiar capability of undergoing development and hatching even without fertilization. Such eggs should be called "unfertilized", not "infertile".

After an egg is laid, it goes through a critical period of time where it is receptive to the fusion of its pronucleus with any one of several pronuclei of sperm that penetrate the egg. If fusion does not occur, the pronucleus of the egg will begin division and establish cell nuclei with only a single set of chromosomes. The differentiated body tissues of drones, however, contain about the same amount of DNA as do those of diploid workers and queens, due to *endomitosis*, the replication of chromosomes within cells without cell division. In fact, most differentiated cells of drones, workers, and queens are *endopolyploid*, that is, they contain several (some more than 16) copies of the chromosome sets.

Sex Determination

Soon after the development of instrumental insemination of honey bee queens, it became apparent that honey bees were subject to extreme deleterious effects of inbreeding. Matings between closely related drones and queens often resulted in a severe loss of brood viability expressed as "shot brood". Shot brood occurs as a consequence of the removal of young larvae from cells leaving a very spotty brood pattern when the cells are capped. Through the work of Drs. Otto Mackensen, Walter Rothenbuhler, and Jerzy Woyke, it was soon determined that shot brood was a consequence of the production of diploid males. Normal, genetically functional males are derived from unfertilized eggs, that is they have one set of 16 chromosomes and are derived directly from the gametes of the queen. Diploid males come from fertilized eggs and, therefore, have two sets of chromosomes, one derived from their mother and one from their father. How can this occur?

The genic mechanism of sex determination involves a major gene, called the sex gene (Fig. 93). Individuals that are heterozygous for this gene, that is they have two different alleles, develop into normal females, queens and workers. Individuals that are homozygous (have two copies of the same allele), or are hemizygous (have only one set of chromosomes, and therefore only one allele) become diploid males. These diploid males are detected by the nurse bees shortly after they hatch from the egg and are consumed. However, they can be raised with some difficulty by removing them from the colony and feeding them royal jelly for 3 days before replacing them in drone cells. After three days, the workers cannot distinguish diploid from haploid males and will raise them normally. Diploid males, though viable, have testes of reduced size and produce diploid sperm that will not produce viable offspring. Therefore, they are undesirable for breeding.

The production of diploid males is inescapable in populations of honey bees because it is dependent on the number of sex alleles and their frequencies. For example, assume there are 10 different sex alleles, each at equal proportions in the population, 1/10th. Any given queen must have two different alleles because she is diploid and female. As a consequence, she shares a sex allele in common with 2/10ths, or 20% of all males with which she may mate (remember that normal males are haploid). Half of the eggs fertilized by sperm from males having a sex allele that

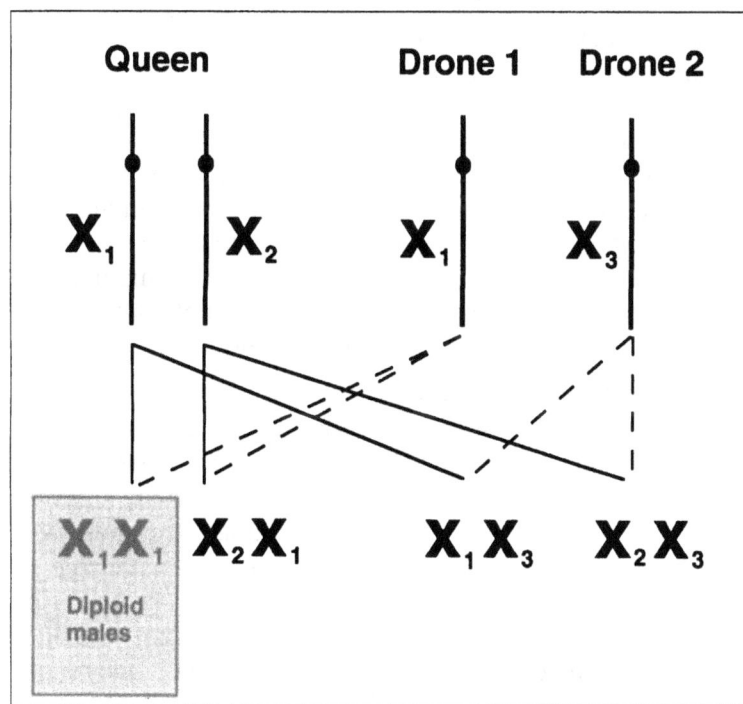

Figure 93. A diagram of sex determination. Lines with solid dots represent the pair of chromosomes that contain the sex locus. One chromosome is inherited from each of the queen's parents. The queen is female, therefore, she must have two different sex alleles designated X_1 and X_2. She is mated to two different drones each with just one chromosome because they are haploid, one of which also has allele X_1 while the other drone has a unique sex allele. The solid lines beneath the chromosomes represent egg gametes of the queen while the dashed lines represent sperm gametes of the drones. The genotypes of the resulting diploid offspring are shown below. The genotype X_1X_1 derived from the mating with drone 1 resulted in the production of a diploid male.

matches hers will develop into diploid males. Therefore, 20% of the males will produce 50% diploid male offspring and on average colonies will lose 10% of their brood from fertilized eggs.

Thelytoky

Thelytoky is the origin of females from unfertilized eggs. The Cape bee of South Africa is well known for the ability of its workers to parthenogenetically produce diploid workers and queens as well as haploid drones. This results from the regeneration of a diploid nucleus soon following meiosis of the egg. Not only is thelytoky of novel interest to geneticists, it provides a valuable tool for the study of honey bee behavior.

The ability of workers of the Cape bee, *Apis mellifera capensis*, to produce females (workers and queens) by parthenogenesis became a topic of great interest and debate early in the twentieth century when G.W. Onions and R.W. Jack reported that females of this race developed from unfertilized eggs laid by workers of queenless colonies. In 1943, Otto Mackensen demonstrated that this phenomenon is not unique to the Cape bee. About 23, 9, and 57% of virgin queens from his stocks of *A. m. caucasica, A. m. ligustica*, and a specially selected "golden" strain, respectively, produced some (less than 1%) workers after they were induced to lay eggs by treatment with carbon dioxide.

Dr. Kenneth Tucker, in 1958, determined that thelytoky occurred most often among progeny of the first brood of a virgin queen. Thelytoky can be induced by causing cessation of oviposition by a queen for a period by confining her and then allowing her to lay again. The origin of the impaternate females was hypothesized by Tucker to be *automictic* where, following meiosis, the egg pronucleus fuses with one of the egg polar bodies and forms a diploid nucleus that undergoes development.

Mosaics and Gynandromorphs

One of the most interesting discoveries during a colony inspection is the occurrence of individuals that are part drone and part worker. These sex-mosaics, called *gynandromorphs*, are not common but do occur occasionally in some colonies. Gynandromorphs can be bilateral with one side male and the other female, or segmental where one end is male and the other female (this is the one that surprises drone collectors!), or individuals can have mosaic patches of different sexes throughout the body. Some gynandromorphs have functional male sex organs and others may have functional female sex organs. These can have special uses in honey bee genetic studies. Walter Rothenbuhler and his associates discovered gynandromorphic bees in research stocks and selected a strain of bees that produced these sex mosaics in much higher than normal frequencies. Using visible mutant marker genes, they found that the primary mechanism leading to gynandromorphs is the result of *polyspermy*—more than one sperm penetrates the egg—with cleavage and subsequent development of an accessory sperm nucleus. This nucleus leads to the production of the male tissue while the zygotic nucleus formed by the union of the egg and sperm pronuclei produces the female tissue. This is not the only mechanism, however. Otto Mackensen reported a case where the female parts were biparental and the male parts were maternal in origin, apparently arising from one of the haploid polar bodies. Tucker found gynandromorphs from unmated queens suggesting union of egg pronuclei to form biallelic female tissue and independent development of a haploid nucleus to generate male tissue.

Same-sex mosaics also occur. These are only detectable if the different tissues derived from different cleavage nuclei contain different visible genetic markers such as eye or integument color mutations. Mosaic males have been found that developed from binucleated eggs while mosaic females have been shown to develop from dizygotic eggs formed from the union of two sperm with two egg nuclei. In 1964, Laidlaw and Tucker reported the most bizarre mechanism of all: the fusion of two accessory sperm to form female tissue while the egg pronucleus fused with a sperm pronucleus (as is normal) and also produced female tissue.

Relationships in the Hive

Queens originate all honey bee genomes. They produce egg gametes with a single set of chromosomes that are either fertilized, and usually develop into females, or are unfertilized and develop into haploid males. As a consequence, queens are genetically both female and male. They function as males because they originate the genomes that are contained in the sperm of their "sons". Males develop from unfertilized eggs and replicate several million times, in their testes, the genome derived from their mother. They change

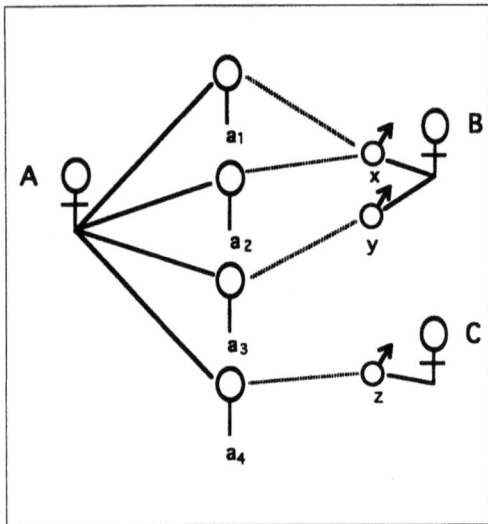

Figure 94. This hypothetical pedigree demonstrates the different relationships that are possible within and among families of honey bees. A-C represent queens. Uncrossed female symbols with lower case letters represent worker offspring. Male symbols designated with lower case letters x-z represent the haploid males. Solid lines represent the egg gametes while dashed lines represent sperm gametes. Individuals a_1 and a_2 are super sisters to each other and full sisters of a_3. a_1, a_2 and a_3 are half sisters of a_4. From Chapter 7, *The Hive and the Honey Bee,* 1992, Dadant and Sons. Used with permission.

the embodying cell of the genome from egg to sperm. Therefore, we can think of queens as both mothers and fathers. Because of this, we advocate the use of genetic, rather than physical, pairing terminology in honey bee genetics and breeding. For example, when we mate drones and virgin queens derived from a single queen, we refer to that as a mother-daughter mating. The physical pairing terminology is a "brother-sister mating". The physical pairing terminology fails to capture the unique characteristics of honey bee genetics and is genetically incorrect. We present several examples of mating systems and proper breeding terminology in the next chapter.

Colonies consist of large numbers of *subfamilies* of workers because queens mate with large numbers of drones (Fig. 94). Members of the same subfamily share both a queen mother and a drone father and are called *super sisters*. We call them super sisters because they inherit exactly the same genes from their father resulting in their sharing 75% of their genes in common by descent compared with only 50% of shared genes among *full sisters* derived from diploid matings. For example, we share only 50% of our genes in common with our brothers and sisters with whom we share both a mother and father. Individuals with different drone fathers belong to different subfamilies and, if the drone fathers are unrelated, the subfamilies share 25% of their genes in common, and the workers of a subfamily are *half sisters* with workers of a different subfamily. *Full sister* relationships are also possible. These occur when brother drones mate with the same queen. In that case, workers with different brother drone fathers are full sisters of each other and share an average of 50% of their genes in common by descent.

Principles of Inheritance

Every honey bee fundamentally resembles every other honey bee. All have basically similar body structures. Within races, more specific resemblance is found. Caucasian bees resemble other Caucasian bees; all Italians have a yellow abdominal marking. The tendency of succeeding generations to resemble the preceding ones is termed *heredity*. But similarity is not total. Italian bees differ in coloration among themselves as well as from the markings of Caucasians. These and other differences are *variations*. The study

Chapter VII

of inheritance in the honey bee is complicated by the wide differences in appearances and functions of the three castes, and by the difficulty in recognizing the same characters in the queen, the drone, and the worker.

The normal eye color in bees is black, but occasionally drones with white eyes appear in a colony of black-eyed bees. The abdominal markings of the drones may also vary and other physical variations may appear. The black eyes, the white eyes, and the abdominal markings are *characters*. The black eyes belong to the same *morphological feature*, eye color, as white eyes, but the two eye colors develop in opposing directions and are spoken of as *contrasting characters*. Black eyes and yellow body color are not contrasting characters; they belong to different morphological features. Characters may be tangible, discernible physical attributes, such as eye colors, or they may be physiological and intangible in nature, such as temper or oviposition rate.

Characters are the end products of the development of the bee from germ cell to adult. Characters result from the action of the *environment* upon determining *factors* in the cells which were transmitted from the parents to the offspring in the germ cells.

The factors are functionally discrete particles, also known as *genes*. The genes are located in the chromosomes, arranged one beside the other like beads on a string. Each gene occupies a particular position in a particular chromosome; this position is the gene *locus*. The two genes of a particular locus of a chromosome pair that are located in the same positions, *loci*, are often called a *gene pair*.

The two genes that occupy corresponding loci of chromosome pairs may be somewhat different, but both affect the same body structure or other feature. For instance, each gene that causes white eye in the bee differs from its corresponding gene that produces black eye, but both are at the same locus. These differing but similar genes of a particular locus are known as *alleles*; one is said to be *allelic* to the other. Each allele contains a different nucleotide sequence that codes for slightly different amino acids that make peptides and proteins that differ slightly in their chemical properties. The white eye gene at a particular locus is thus an allele of the black eye gene at the same locus. Genes affecting a given feature are not necessarily allelic. The four known genes which produce white eyes, for example, are at different loci and are non-allelic to each other.

When the two alleles are present in the cells, one of them often appears to have a stronger influence on the trait they both affect. Thus, when the cells of the bee have both a black eye gene and a white eye gene, the bee's eyes are black. The black eye gene is said to have *dominance*, for it "dominates" the white eye gene, the *recessive* gene. Only when both genes of a pair are recessive does a recessive character develop. Thus, when two white eye allelic genes are present in the cells of female bees, the bees have white eyes; but when one gene is white eye and the other black eye, the bees have black eyes. Alleles can also be *codominant*, meaning that both alleles are expressed. For instance the snow gene locus has three alleles for eye color. The snow allele results in white eyes when *homozygous* (both alleles are the same); the tan allele of the snow locus results in tan colored eyes when homozygous. A worker or queen has red colored eyes when she has one tan and one snow allele (*heterozygous*). Both the snow and tan alleles are recessive to the *wild type* black eye color allele.

When the two genes of a gene pair in the female bee are alike, they are said to be in *homozygous* condition, and the bee is homozygous for the character produced by the genes. The bee then must breed "true" for the particular character, for all germ cells it produces have the same gene for the character. But when the two genes of a pair are unlike (in *heterozygous* condition), the bee does not breed true for the character; half of the gametes receive each kind of gene.

Drones, having only one set of chromosomes, have only one of the genes of a gene pair. At a particular locus, a drone may have a black eye gene or a white eye gene, but not both. The single gene determines the character. The drone is, in effect, homozygous for all genes. And all spermatozoa produced by a particular drone must also have identical genes. Such haploid individuals are said to be *hemizygous*.

As mentioned earlier, the normal eye color of bees is black, but sometimes a drone is found with eyes which are white, chartreuse, or red; or there may be no eyes at all. These deviations from normal are frequently heritable and usually arise through a change in a normal gene, or one of its alleles. When a gene changes and the resultant gene produces an effect on a trait different from the effect produced by the original gene, the change is called *mutation*. Usually the mutant gene is recessive to the normal gene.

When more than one kind of gene arises at a particular locus by mutation, there are more than two alleles though only two can be in a cell at the same time, and an individual female bee can have only two of them, and a drone only one of them. In these cases the normal gene and the mutant genes form a series of alleles which are called *multiple alleles*. This is illustrated by the allelic series at the chartreuse locus in bees: ch^+ (normal black), ch^1 (chartreuse-1), ch^2 (chartreuse-2), ch^c (cherry), and ch^r (red).

The two alleles, the normal one for black eye and the mutant one for white eye, exist together in the same cells of a queen heterozygous for them. The black eye gene is dominant over the white eye gene, and the queen has black eyes. But the black eye gene does not affect the white eye gene itself. Alleles affect only the characters, not each other. When the eye color alleles separate at meiosis, during the maturation of the egg, one remains in the egg pronucleus and the other is eliminated in the polar body. It is a matter of chance which of the two eye color genes remains in the egg pronucleus after the meiotic divisions are finished, but on the average each allele will get into the pronucleus an equal number of times. Thus, approximately half of the eggs will have the black and approximately half the white eye allele. If these eggs are unfertilized, they will develop into drones in the ratio of about one black-eyed drone to one white-eyed drone. The separation of the alleles at meiosis and their distribution to different gametes, where each may again assert its effect on a character, is called *segregation*. Segregation is a constant feature of all diploid inheritance (Figs. 89 and 90). The genes on homologous chromosomes segregate and are distributed to germ cells independently of genes on other chromosomes.

Each chromosome of a cell has many genes, each of which exerts a more or less pronounced effect upon respective traits. The gene for black eye, for instance, may be on the same chromosome as a dominant gene affecting some unrelated character. Since the two genes are on the *same chromosome*, they tend to stay together when the gametes are formed and are said to be *linked*. When homologous chromosomes come together at meiosis, they exchange segments, as *crossing over*, and a certain percentage of the time the point of exchange occurs between these two particular linked genes. They thus become separated and go into different gametes in which they have new linkage relationships which are retained until crossing over again takes place.

The paired genes of an organism make up the *genotype*; the characters due to these genes make up the *phenotype*. A single set of chromosomes as occurs in unfertilized egg and sperm cells is called a *genome*. The term genome is also used to describe the collection of different genes that constitute an individual. Geneticists combine genomes through controlled matings of queens and drones to produce specific genotypes for genes that are important to them.

Polygenic Traits

It is assumed that many characteristics of honey bees are *polygenic*, that is they are determined by many genes, each having a very small effect. This is in contrast with the case of sex determination and eye color discussed above where single genes have major effects on the phenotype. Polygenic traits usually demonstrate continuous variation and must be measured in some way to define different phenotypes. These are called *quantitative traits*. A gene that affects a trait that varies in a measurable way is called a *quantitative trait locus* (QTL). Each QTL may affect a trait differently: some may have major effects while others are minor. Genes with relatively minor effects are sometimes called *modifier genes*. Some QTLs may be recessive, others codominant. Little is known about the number and action of genes that are responsible for quantitative traits, however, new genetic techniques recently have been developed that help determine the numbers and action of QTLs.

Genetic Maps

Advances in biotechnology and molecular genetics have provided new tools for studying the organization of DNA in honey bees. These tools facilitate the manifestation and visualization of molecular DNA *markers* that are scattered throughout the genome. DNA markers are regions of the genome that vary in nucleotide sequences in a way that they can be detected.

It is beyond the scope of this book to describe the techniques used to produce and visualize DNA markers, however, these markers have been used to characterize the honey bee genome and to construct *genomic maps* of honey bee genes (Fig. 95). DNA markers, like genes, are organized into chromosomes that segregate together. Crossing-over breaks stands of DNA and can exchange both genes

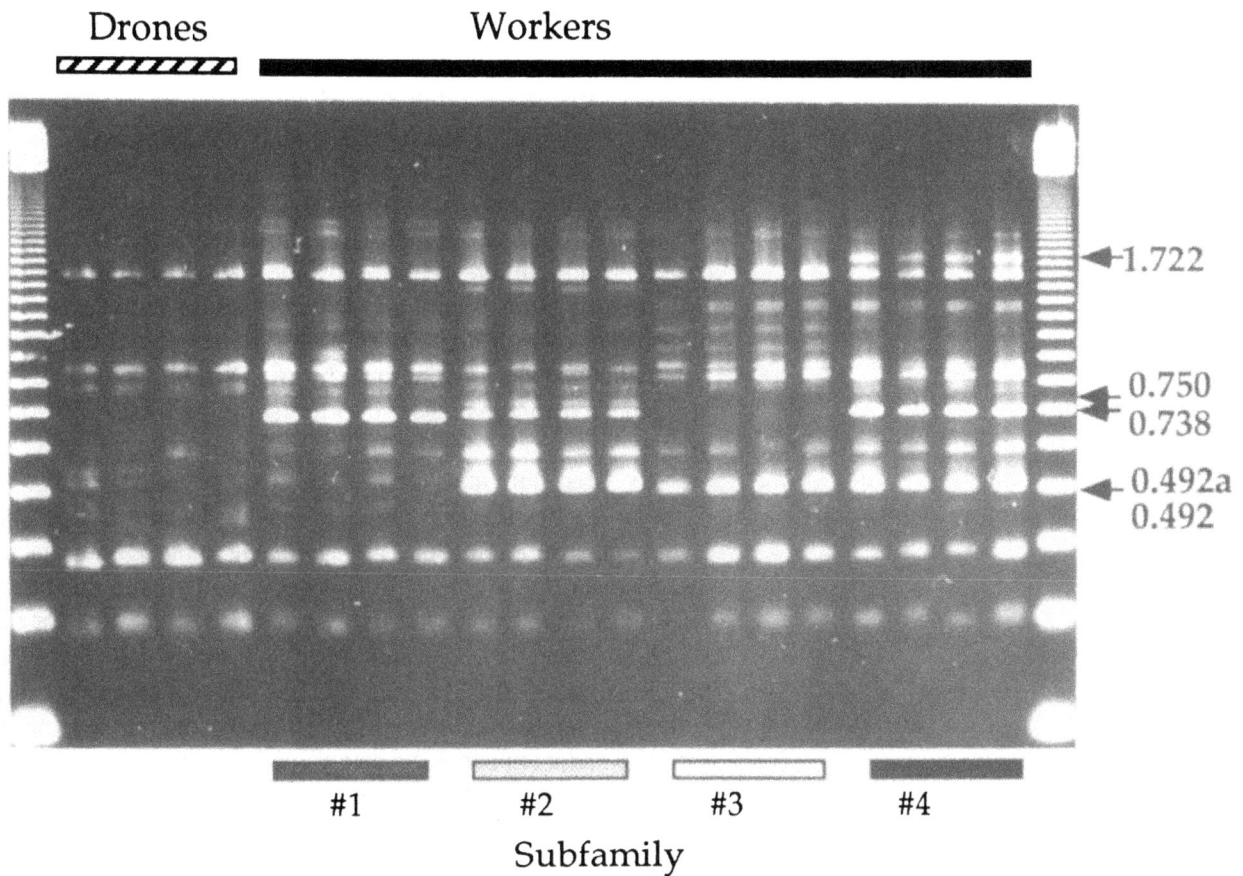

Figure 95. Photograph of an agarose gel stained with ethidium bromide demonstrating random amplified polymorphic DNA (RAPD) markers. DNA acts as a template for the synthesis of duplicate DNA copies. DNA was extracted from drones and workers and DNA copy fragments were synthesized in a tube. The template to make the fragment was present in the genetic code for some individuals, but not in others, resulting in a variable genetic marker. For instance, each vertical "lane" contains DNA fragments synthesized from single individuals. (The bands in the first and last lanes contain a DNA size marker that is used to determine the sizes of the fragments.) The four lanes on the left (lanes 2-5) are from drones of the mother queen of the 17 workers shown to the right of the drones. The queen was mated to 4 different drones producing subfamilies 1-4. The marker designated 1.722 at the right margin of the figure is not present in the drones of the queen, therefore, the queen does not have that DNA template, but it is present in the last 4 workers. Those four workers had the same father and are members of subfamily 4. Therefore, they inherited the template to make that marker from their father. None of the other 3 subfamilies had that marker. Likewise, marker .492a was found in workers from subfamilies 2-4 but not in subfamily 1. Therefore, the queen mother did not have that DNA template but her mates 2-4 did. From Fondrk, Page and Hunt. 1993. *Naturswissenschaften* 80:226-231. Copyright Springer-Verlag, used with permission.

and markers between chromosomes (see figure 89). If two genes, two markers, or a gene and a marker are close together on a chromosome, it is unlikely that a crossover will occur between them during meiosis and they will usually remain together in the genome of a new gamete. The farther apart they are on the chromosome, the more likely they will be recombined into different chromosomes and segregate into different gametes. The association of genes and markers on chromosomes is called *genetic linkage*. The amount of recombination between genes and markers is a measure of their distance from each other on the chromosome. For instance, genes on different chromosomes are not linked and are expected to *cosegregate* into the same genomes 50% of the time.

Drs. Greg Hunt and Robert Page constructed a genomic map of the honey bee that consists of 365 DNA markers mapped onto 26 linkage groups (Fig. 96). They were able to place onto this map the gene for sex determination, a gene responsible for the production of an enzyme, *malate dehydrogenase*, and a major gene for black body color. In addition to these single gene traits, they have also mapped QTLs that affect whether a foraging worker collects pollen. Variation among workers in pollen and nectar foraging behavior appears to be determined in large part by two major genes, and probably some modifiers.

Hybrid Vigor

It has long been known that when you cross parents from different inbred lines or from different races or strains that the offspring can have characteristics that are superior to both of the parents. This phenomenon is called *hybrid vigor*. Hybrid vigor, also called *heterosis*, may occur as a consequence of dominance at gene loci. Dominance can be of three types. When you cross a queen with white eyes to a black-eyed drone, all worker progeny resemble the father, showing *complete dominance* for the black eye-color gene. Dominance for some characteristics may be *incomplete*

Figure 96. (Facing page). A linkage map of the honey bee based on DNA markers. Only 4 of the major linkage groups are shown. The numbers to the left of the bars show the amount of recombination that occurs between the marker loci shown on the right. For instance, on linkage group I crossing-over occurs between marker loci Q9-38f and 365-1.98 in about 38.2% of meiotic cell divisions. Note that the sex locus (X) is shown at the bottom of linkage group III. From Hunt and Page, 1995. *Genetics* 139: 1371-1382.

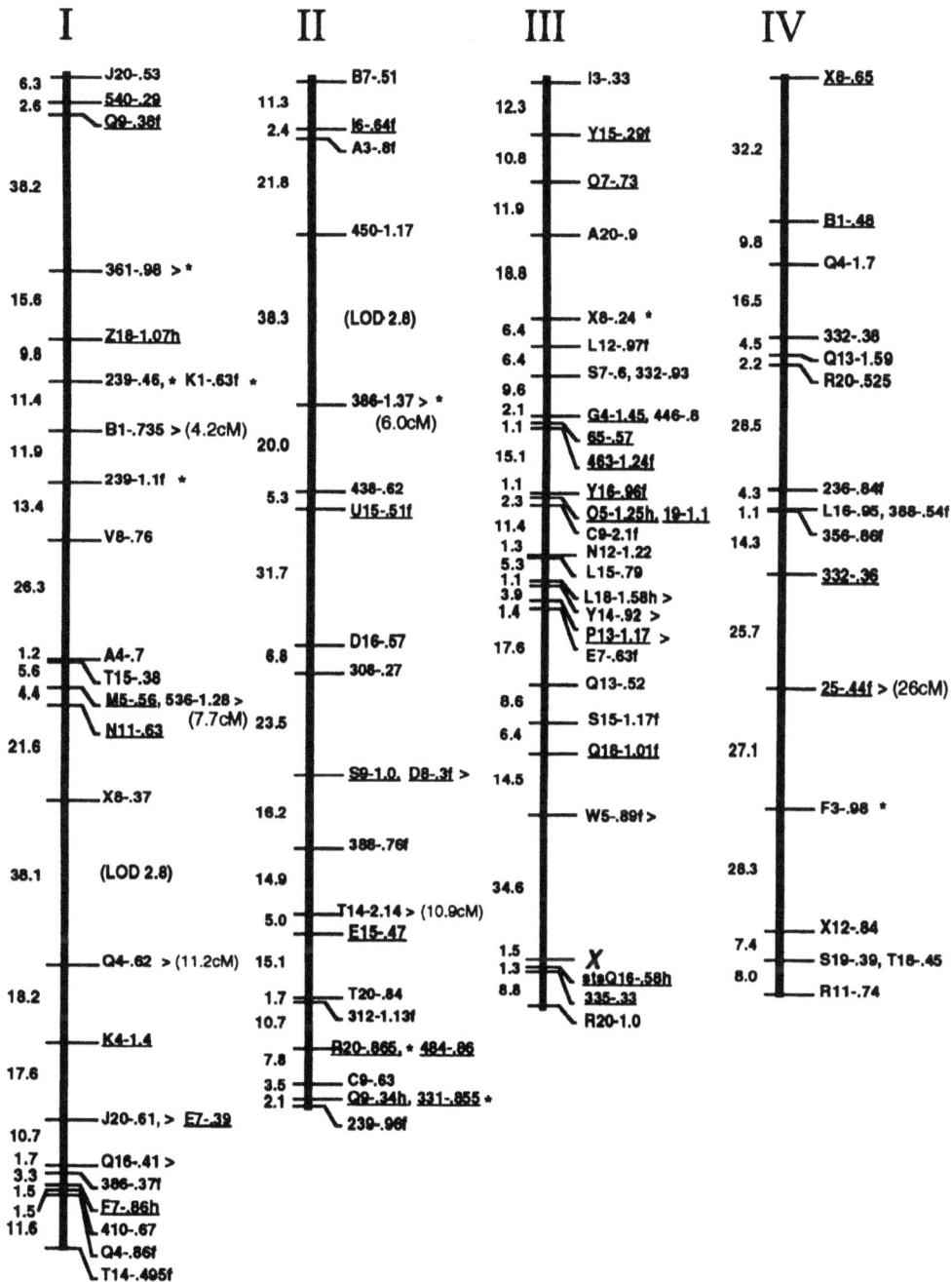

I

6.3 — J20-.53
2.6 — 540-.29
— Q9-.36f

38.2

— 361-.98 > *

15.6
— Z18-1.07h
9.8
— 239-.46, * K1-.63f *
11.4
— B1-.735 > (4.2cM)
11.9
— 239-1.1f *
13.4
— V8-.76

26.3

1.2 — A4-.7
5.6 — T15-.38
4.4 — M5-.56, 536-1.28 >
         (7.7cM)
— N11-.63
21.6

— X8-.37

38.1   (LOD 2.8)

— Q4-.62 > (11.2cM)
18.2

— K4-1.4
17.6

— J20-.61, > E7-.39
10.7
1.7 — Q16-.41 >
3.3 — 386-.37f
1.5 — F7-.86h
1.5 — 410-.67
11.6 — Q4-.86f
— T14-.495f

II

— B7-.51
11.3
2.4 — I6-.64f
— A3-.8f
21.8

— 450-1.17

38.3   (LOD 2.8)

— 386-1.37 > *
         (6.0cM)
20.0

5.3 — 438-.62
— U15-.51f

31.7

6.8 — D16-.57
— 308-.27

23.5

— S9-1.0, D8-.3f >

16.2

— 388-.76f

14.9
5.0 — T14-2.14 > (10.9cM)
— E15-.47
15.1

1.7 — T20-.84
10.7 — 312-1.13f

7.8 — R20-.865, * 484-.86
3.5 — C9-.63
2.1 — Q9-.34h, 331-.855 *
— 239-.96f

III

— I3-.33
12.3
— Y15-.29f
10.8
— Q7-.73
11.9
— A20-.9
18.8

6.4 — X8-.24 *
6.4 — L12-.97f
9.6 — S7-.6, 332-.93
2.1 — G4-1.45, 446-.8
1.1 — 65-.57
15.1 — 463-1.24f
1.1 — Y16-.96f
2.3 — Q5-1.25h, 19-1.1
11.4 — C9-2.1f
1.3 — N12-1.22
5.3 — L15-.79
1.1 — L18-1.58h >
3.9 — Y14-.92 >
1.4 — P13-1.17 >
17.6 — E7-.63f

— Q13-.52
8.6
— S15-1.17f
6.4
— Q18-1.01f
14.5

— W5-.89f >

34.6

1.5 — X
1.3 — stsQ16-.58h
8.8 — 335-.33
— R20-1.0

IV

— X8-.65
32.2

— B1-.48
9.8
— Q4-1.7
16.5
4.5 — 332-.38
2.2 — Q13-1.59
— R20-.525
28.5

4.3 — 236-.84f
1.1 — L16-.95, 388-.54f
14.3 — 356-.86f

— 332-.36
25.7

— 25-.44f > (26cM)
27.1

— F3-.98 *
28.3

7.4 — X12-.84
8.0 — S19-.39, T18-.45
— R11-.74

where the offspring are intermediate between the parents for a character determined by one gene, but strongly favor the parent carrying the dominant gene. *Overdominance* (or *underdominance*) may occur when individuals that are heterozygous (have two different alleles at a gene locus) are more (or less) extreme than either homozygote (have two of the same allele) for that character. Overdominance can lead to hybrid vigor.

Hybrid vigor is most often believed to be a consequence of directional dominance at multiple loci of a polygenic trait. For example, assume that the size of a worker is a polygenic, quantitative trait with four QTLs, A, B, C, and D (this is just a hypothetical example, the actual genetics of size are unknown). Assume that each QTL can increase the size of a worker by an equal amount and that for each QTL there is a dominant and a recessive allele with A, B, C, and D dominant over a, b, c, and d, respectively. In that case, genotypes AABBCCDD and AaBbCcDd would both yield the largest possible phenotype while genotype aabbccdd would yield the smallest. Assume that strain I has a fixed genotype of AAbbCCdd and strain II has a fixed genotype of aaBBccDD. The strains would be the same size, intermediate between the possible size extremes. Strain I gametes would all contain a AbCd genome while gametes of strain II would all be aBcD. If these strains were crossed, all progeny produced would be F_1 hybrids, would have AaBbCcDd genotypes, and would be two "units" larger than either parental strain. They would exhibit hybrid vigor.

SELECTIVE BREEDING

The beekeeper would like a simple way to improve his stock and no doubt would prefer to have little or nothing to do with chromosomes, genes, and the relationships of his queens. Unfortunately, there is no simple formula for bee breeding which will always work and which will entail little effort on the part of the beekeeper. The beekeeper can improve stock, with little knowledge of animal breeding, if he is willing to make the effort that is required. The beekeeper must be willing to make careful observations on each colony that is tested and must provide an abundance of selected drones to ensure that a good proportion of his queens are mated with the desired stock.

Stock Evaluations

A colony of bees consists of two generations: the queen who is the mother of the colony, and the workers that are her offspring. In judging a colony, both generations must be evaluated. The queen is judged by the quantity of brood she produces and whether she misses many cells when she lays eggs (Fig. 97). A good queen will lay 1200 or more eggs each 24 hours during the intensive brood-rearing time of year. She will not ordinarily have more than three to five empty cells per hundred among the sealed brood of the better combs. She will lay to the bottom of the combs and as far to the sides and top as the stores of pollen and honey permit (Fig. 98). During the initial period of expansion of the brood nest, she will lay as many eggs as the bees can possibly care for, and often she seems to lay more than they can care for. She will maintain a high egg-laying rate until she is restricted by a lack of room, or until the bees change her diet. The purity, but not necessarily the quality, of the queen's own breeding is indicated by the evenness of the color of her drones, as well as by her own appearance.

The selection of breeder queens from each generation is a critical part of a breeding program. Any progress to be made in improving the stock will depend largely on these selections. Records should be kept for each colony tested. They should include observations on the amount of surplus honey produced by

Figure 97. A good comb of brood.

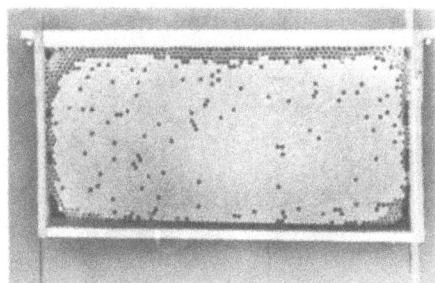

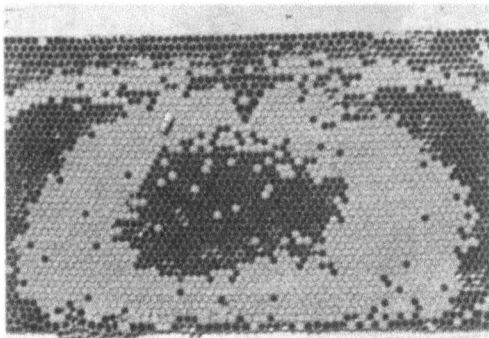

Figure 98. Emergence pattern of brood where the eggs had been laid first at the center of the comb and then in rings to the comb edges.

the colony, swarming tendency, gentleness of the bees, their quietness on the combs, solidness of the brood, quantity of the brood, resistance to diseases, and willingness of the bees to draw foundation. Evenness of abdominal markings is not too important at first, although a standard should be established towards which to work. Evenness of abdominal markings is a measure of the purity of breeding. The best colonies which also have color markings closest to the goal should be chosen as breeding colonies. After several generations of careful selection, the color markings of the queens, bees, and drones should become quite uniform.

The testing of potential breeding stock is subject to many errors which must be guarded against. Drifting of the worker bees is one of the most serious. Test colonies should be located so that drifting is reduced to a minimum. Colonies should not be arranged in even rows nor bunched together. They should be located several feet apart and with natural markers for each colony.

Differences in the starting strength of the colonies and variations in colony management may introduce errors. All colonies under test should either be placed in one yard or representative colonies from each line should be distributed among several testing yards. Locations vary even within short distances, therefore, location differences must be taken into account in evaluating the stocks.

Principles of Breeding

The goal in bee breeding is to concentrate and intensify desirable characters in a strain or line and at the same time eliminate the undesirable ones. Since characters are expressions of their genes, this means concentrating in a line a maximum number of genes which contribute toward the desired characters. In beginning a stock improvement program, the bee breeder must decide which characters are the most important, and must confine his or her efforts to these. It is relatively easy to establish one character in a line if the character is dependent upon a single gene pair. But as more characters are combined, the problem rapidly becomes more complex. It becomes especially difficult when the various characters are influenced by several gene pairs and multiple alleles.

Characters may conveniently be classified into two general types: those that either are fully developed or do not appear at all, *qualitative characters*, and those that appear in varying degrees and require measurements to distinguish them, *quantitative characters*. Eye color illustrates the first type. The yellow coloring of the abdomen illustrates the second type. Most of the economically important characters of the bee are believed to be quantitative.

The expression of qualitative characters is often governed by alternative alleles at a single gene locus. This is the situation in the inheritance and expression of white eye color in the bee. Here one of the alleles is dominant over the other. In some cases, neither gene may be dominant, but the two alleles may produce an effect intermediate between the two extremes. In still other cases the dominant allele of any one of two or more factor pairs may produce a particular character. For instance if two gene pairs such as *Aa* and *Bb* have the same effect on a character, the character will appear fully developed even if only one dominant is present, as *Aabb*, or *aaBb*. If no dominants are present, as *aabb*, the opposite or contrasting character is produced.

Quantitative characters are often dependent upon more than one gene locus. Quantitative genes (QTLs) exhibit a cumulative effect so that two similar genes from the same gene pair or from different gene pairs for a particular character produce a greater effect than one gene. All factors do not necessarily contribute equally to a character, as in the case of genetic control of color variation. According to Dr. Jerzy Woyke, a Polish honey bee geneticist, the color of the abdomen of bees is under the control of a single gene with major qualitative effects and several modifier genes, each of which have small quantitative effects. The major gene has three alleles, Y, y^{bl}, and y^{bc}. y^{bl} is recessive to Y and results in dark colored bees when homozygous, y^{bl}/y^{bl}. Y/Y and Y/y^{bl} workers, queens, and drones maintain their yellow color patterns. African bees have a third allele, y^{bc}, that turns the color brown, but only in drones. An additional 6 or 7 genes with minor effects alter the degree of black and yellow. For instance a worker that is y^{bl}/y^{bl} that has yellow-coding alleles at these minor loci will be somewhat more yellow, or less dark, than one that has all black coding alleles. The same holds for one that is Y/Y, she would be less yellow if she had black-coding alleles at the minor loci but not as a dark as a y^{bl}/y^{bl} worker with yellow coding alleles.

Expectations

Expectations for breeding programs should be realistic. The elimination of disease problems, or having colonies with 1,000 pound average honey yields are not realistic objectives. Instead, a reduction in incidence of disease resulting in a reduction in the need to chemically treat hives, or a modest increase in honey production, may be realistic. Improvements in apicultural practices will often have greater effects than selective breeding for economically important traits. For example, good hygienic practices of beekeepers will reduce the spread of American foulbrood (AFB) and may have a greater effect on the incidence of disease than selecting for resistance. Requeening regularly should have significant effects on colony performance and honey production.

Selection is a necessary, continuous process in order to produce and maintain improved stocks. It is unrealistic to expect to maintain stocks with selected characteristics without continual selection. Selective progress will begin to deteriorate as soon as selection is relaxed on the breeding population.

The Assay

You must have a sufficiently reliable assay before you can select for any trait. An assay is simply a method to categorize or quantify the variability that you observe for the characteristic you wish to improve. The most important consideration for a good assay is to control the environment in which the assay is performed. That usually requires having the best apicultural practices possible, before you begin selecting. Assays should have a high degree of repeatability under a given set of environmental conditions. Examples of assays are: 1) the short-term gain in weight of colonies as a measure of honey production, 2) the measured area of stored pollen in a colony as a measure of pollen collecting and storing behavior, 3) the time it takes a colony to uncap and remove a specific number of cells of diseased or freeze-killed larvae, a test used for evaluating resistance to AFB and chalkbrood disease; and 4) the number of adult tracheal mites found in the tracheae of young workers five days after they are introduced into tracheal-mite infested colonies, a measure of tracheal mite resistance.

Assays can be conducted in field hives, nucleus colonies, or in the laboratory. Environmental conditions are more easily standardized in laboratory tests. However, results from selection using laboratory tests or nucleus colonies should be verified under commercial field conditions. It should be kept in mind that you cannot directly select for resistance to diseases and parasites that you don't have. You also cannot directly select for resistance while you chemically treat for the disease, in that case the disease or parasite can no longer act as a selective agent. You may, however, select directly for known mechanisms, such as hygienic behavior. The results from the selection must be tested with diseased or parasitized colonies because presumed mechanisms selected for resistance under constrained conditions may not have the same effect on full-sized hives in commercial apiaries.

fundamentals

Selective breeding requires four simple steps. 1) First, colonies, or individual workers, drones, or queens are evaluated in some base population from which you will select your breeding stock. Then, 2) queens of colonies having the desired characteristics are selected as breeders for the next generation. 3) Matings are controlled between queens and drones raised from the selected breeders. 4) Progeny of the new queens are evaluated, either as individuals or as a colony. The progeny will resemble their parents and selective improvement will be obtained if a sufficient proportion of the observed trait variation in the parental population was due to genetic differences. Hence, selective breeding is a process of measurement, and trial and error.

Natural selection may also produce disease resistant populations of honey bees. Populations that have been exposed to severe diseases for prolonged periods of time should become resistant relative to populations that have not been exposed. For instance, the Asian honey bee, *Apis cerana,* has had a long term exposure to varroa in Asia and is relatively resistant to severe damage by this parasite. American foulbrood disease devastated the beekeeping industry of Hawaii in the 1930s resulting in its near total collapse. However, by the late 1940s AFB was a less severe problem in larvae from Hawaii and Vic Thompson and Walter Rothenbuhler at the Ohio State University were able to demonstrate resistance to infection to AHB spores of larvae from some colonies.

Comparative Evaluation

It is necessary to evaluate stock that has been produced through selection. There are numerous unsubstantiated claims proliferating in the beekeeping trade journals of disease resistant stocks. It is important to have a benchmark against which selective progress can be compared in order to separate the effects of the selection program from effects of a changing evaluation environment. The benchmark for comparison should be another population of bees that have not been subjected to the same selection.

Three methods of comparison are commonly used. 1) Comparison with a "standard" stock that remains genetically unchanged throughout the time that the selection is occurring. This is very difficult with honey bees because it requires either the maintenance of inbred lines that are homozygous at all loci (*isogenic*), and, therefore, genetically constant through time, or it requires the long term survival of a set of queens whose progeny can be used for comparison with subsequent generations of the selected population. 2) Conduct two-way selection. Two-way selection requires maintaining two selected populations. One is selected for higher values of the selected trait, while the other is selected for lower values. Its advantage is that it quickly allows the determination of the genetic basis of observed phenotypic differences within a breeding population. However, its disadvantages are that it requires a double effort by maintaining commercially worthless low-trait colonies, and it still requires additional evaluation of the high-trait population against other unselected stocks. Without the additional evaluation, it is impossible to determine if only the low trait colonies changed with selection, only the high, or both changed. 3) Comparison with a "random" stock that represents the unselected gene pool from which the original breeder queens were selected to initiate the breeding program.

Sources of Bees

An important question to ask before beginning a breeding program is "where do I look for stock?" There are 4 potential sources.

Local Commercial Bees

Locally available commercial stocks probably will have genetic variability for most characteristics of commercial importance, if it exists anywhere. Local bees are also likely to be better adapted to the local conditions than bees from other areas.

Commercial Bees From Afflicted Areas

Commercial bees from areas that have an historical association with a specific disease may constitute the foundation population for breeding for resistance. These bees may already have been subjected to artificial or natural selection and already demonstrate some genetically-based resistance or tolerance.

Races of Bees

Specific races of bees may be imported and tested for commercially important characters. Perhaps they have brood rearing or honey production characteristics that are of interest to a breeder. Or, perhaps they have a long-time association with a particular disease and are already naturally resistant. Such is the claim for the Carniolan "Yugo" bees imported into the United States by the U.S. Department of Agriculture for their presumed resistance to varroa and tracheal mites. Or, as is the case with Africanized honey bee resistance to varroa in the New World, imported bees may be naturally resistant. Alternatively, imported bees may be no better than the local commercial bees but when they are crossed with local commercial stocks their progeny may demonstrate hybrid vigor resulting in increased honey production or reduced disease symptoms. However, bee breeders should be careful that they are not importing objectionable characteristics along with the desirable traits, like the strong defensive behavior and swarming tendencies of the African honey bees imported into Brazil in 1956.

Feral Bees

Feral bees may be used as a reservoir for genetic diversity and may potentially have naturally-selected characteristics. The feral bees of California demonstrate local genetic differentiation for the enzyme malate dehydrogenase, suggesting that they have undergone local adaptation and, therefore, are probably genetically differentiated from commercial honey bees. California feral bees also vary in size throughout their geographical range with larger bees located to the north and smaller bees to the south and in the dryer, desert regions. More than 85% of all feral colonies in some areas of California were killed by varroa between 1990 and 1994. Therefore, it is possible that strong natural selection on feral colonies, along with their, at least partial, genetic isolation from commercial bees, has resulted in some degree of disease resistance in feral bees. However, it is surprising that there are no research studies of naturally nesting, feral bees to determine levels of resistance to diseases.

Mating Designs

Breeding Terminology

The genetic system of honey bees is different from that of other commercially important plant and animal stocks because of its method of sex determination (discussed in Chapter VII). Males are derived directly from unfertilized eggs of queens (and sometimes workers) and have no father. As a consequence, queens are the source of all *genomes* (single sets of chromosomes contained in sex cells called *gametes*). Males have only the single set of chromosomes they received from their mother and each cell of the male contains only the genes it inherited from her. In the testes of the male, this single set of chromosomes is replicated 6-10 million times and packaged in sex cells called spermatozoa. Every sperm cell contains the same genome that came directly from the queen. Because of this, queens effectively function as both female and male. They function as a female through their production of eggs that are fertilized and develop into females (and sometimes diploid males, see Chapter VII). They function as males through the replication of the genome they contribute to the sperm of their "sons". Males function only to replicate the genome of the queen and to change the embodying sex cell from egg to sperm.

Two types of terminology have been used in the honey bee breeding literature to describe mating systems. One is based on the *physical pairing* of parents, the other, *genetic pairing*, is based on the origins of the genomes that combine as a consequence of mating. Physical pairing terminology does not reflect the unique characteristics of honey bee genetics and is imprecise. With genetic pairing terminology, queens act as both the male and female parent because they originate all genomes. To illustrate the differences, suppose that a virgin queen is mated to a drone produced by her own mother. With physical pairing terminology this would be considered a *brother-sister* mating. However, with genetic pairing terminology the mating is *mother-daughter*. Two fe-

Table 2. Comparison of physical and genetic pairing terminology for selected crosses used in honey bee breeding.

Type Cross	
Physical pairing	Genetic pairing
mother-son	self
sister-brother	mother-daughter
aunt-nephew	sister-sister
uncle-niece	grandmother-granddaughter

males that share the same queen mother and the same drone father are called *full sisters* with physical pairing terminology but they are *super sisters* with genetic pairing. They are called super sisters because they all inherit exactly the same set of genes from their father, resulting in their sharing a higher proportion of their genes in common (75%) compared with full sisters (50%). With diploid genetics, each full sister inherits a different set of genes from both the mother and the father because both parents undergo successful meiosis when they produce gametes and each gamete contains a different genome as a result of crossing over and independent assortment of chromosomes. Tables 2 and 3 demonstrate the differences between physical and genetic pairing terminology for some specific crosses.

Inbred-Hybrid Breeding

After the development of instrumental insemination technology in the 1930's and 1940's, honey bee geneticists were eager to begin controlled mating breeding programs that had proven so successful for improving other plant and animal stocks. In particular, they wanted to join in the great success of corn breeders who had developed superior hybrid varieties by first inbreeding, then crossing inbred lines. Inbred-hybrid breeding programs take advantage of hybrid vigor (discussed in Chapter VII) and have resulted in improved stocks of honey bees such as the Starline™ and Midnite™ stocks produced by Dr. Bud Cale when he worked for Dadant and Sons, Inc., Hamilton, Illinois.

Inbred-hybrid breeding begins by developing inbred lines. Several mating systems may be employed during inbreeding, but two considerations should be kept in mind: 1) the inbreeding rate,

Table 3. Comparison of physical and genetic pairing terminology for relationships between related workers or queens.

| Description | Terminology | | Diploid equivalent | Proportion genes shared |
	Physical	Genetic		
share mother and drone father	full sister	super sister	none	0.75
share mother but different "brother" fathers	none	full sister	full sister	0.50
share same drone father but different mothers*	paternal half sister	super half sister	none	0.50
share mother only	maternal half sister	maternal half sister	maternal half sister	0.25
share "paternal" grandmother	cousin	paternal half sister	paternal half sister	0.25

This can occur with instrumental insemination when the semen of single drone is partitioned between different queens.

and 2) the generation time. The rate of inbreeding refers to the rate at which genetic variability is lost from the inbred lines. For example, if you chose to use selfing as your mating system, your inbred line will lose 50% of its remaining genetic variability each generation. If you began with a queen that was heterozygous (had different alleles) at half of her gene loci, queens of the 3 subsequent generations would have 25%, 12.5%, and 6.25% of their loci heterozygous, respectively. The generation time would be the total time needed to produce a queen (16 days from the time the egg is laid), plus the number of days after the adult queen emerges, before insemination (8 days), plus the time it takes her to begin laying eggs in drone-sized cells (5 days *or more*), plus the development time for drones from egg to adult (24 days), and the time it takes the drones to become sexually mature (10 days). There-

Table 4. Inbreeding coefficients for generations of regular, closed mating systems.

Selfing	$F_t = 1/2 \, (1 + F_{t-1})$
Maternal mother-daughter	$F_t = 1/4 \, (1 + 2F_{t-1} + F_{t-2})$
Paternal mother-daughter	$F_t = 1/2 \, (1 + F_{t-1})$
Super sister-super sister	$F_t = 1/8 \, (3 + 4F_{t-1} + F_{t-2})$
Half sisters (drone mothers super sibs)	$F_t = 1/32 \, (7 + 16F_{t-1} + 8F_{t-2} + F_{t-3})$
Half sisters (drone mothers full sibs)	$F_t = 1/16 \, (3 + 8 F_{t-1} + 4 F_{t-2} + F_{t-3})$
Half sisters (drone mothers half sibs)	$F_t = 1/8 \, (1 + 6F_{t-1} + F_{t-2})$
Aunt-niece (single drone)	$F_t = 1/16 \, (3 + 8 F_{t-1} + 4 F_{t-2} + F_{t-3})$
Aunt-niece (multiple drones)	$F_t = 1/8 \, (1 + 4F_{t-1} + 2 F_{t-2} + F_{t-3})$

fore a *minimum* of 63 days are needed for each generation. This is a conservative estimate of time because young queens frequently resist laying unfertilized eggs that develop into males. On the other hand, mother-daughter mating results in a slower rate of inbreeding (Fig. 99) but the generation time is faster. The goal of the inbreeding phase of inbred-hybrid breeding is to get each line to be homozygous for all variable gene loci.

The loss of heterozygosity in a line is usually expressed as an inbreeding coefficient which is the proportion of the originally heterozygous loci that have become homozygous as a consequence of inbreeding. This relationship can be expressed by the formula: $H_t = H_o(1 - F_t)$ where H_t = the proportion of remaining heterozygous loci in some given generation, t; H_o = the initial propor-

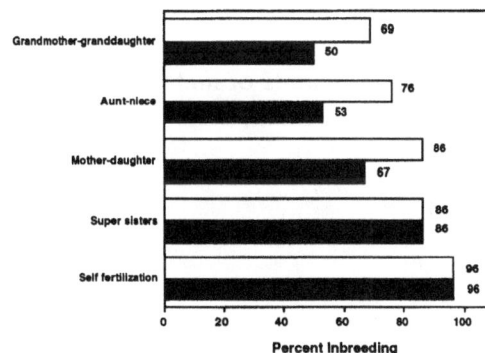

Figure 99. This figure compares the degree of inbreeding within closed mating systems after five generations (open bars) with the amount of inbreeding expected in twelve months (solid bars). The differences occur because of the varying amount of time required per generation with the different mating schemes. This figure was redrawn from Polhemus and Park 1951. *Journal of Economic Entomology*, 44: 639-642.

Selective Breeding

tion of variable gene loci that were heterozygous in the line; and F_t = the inbreeding coefficient in generation t. It is usually assumed that any individual in a randomly mating population, e.g. no inbreeding, is heterozygous at about half of the gene loci that have variable alleles ($H_o = 0.50$). Another assumption is that there are two alleles at each variable locus and that the alleles are at equal frequencies of 50%. Table 4 gives equations for calculating the inbreeding coefficient for regular systems of inbreeding and Table 5 presents inbreeding coefficients for up to 20 generations with regular mating designs. However, sometimes irregular mating systems are employed and inbreeding coefficients are calculated directly from a pedigree.

Calculating Inbreeding Coefficients from Pedigrees

Calculating an inbreeding coefficient is not something that many bee breeders will ever have to do, however, knowing how they are calculated may lead to a better understanding of what they are. An inbreeding coefficient is the probability that both alleles at a given gene locus are identical because they came from the

Table 5. Coefficients of inbreeding for selected, regular mating designs

Mating	Generation						
	1	2	3	4	5	10	20
Selfing	.500	.750	.875	.938	.969	.999	1.00
Mother-daughter	.250	.375	.500	.594	.672	.886	.986
Father-daughter	.250	.375	.438	.469	.484	.500	.500
Super sisters	.375	.562	.703	.797	.861	.979	1.00
Full sisters	.250	.375	.500	.594	.672	.886	.986
Aunt-niece (single drone)	.188	.281	.375	.457	.527	.764	.942
Grandmother- granddaughter	.125	.250	.281	.359	.430	.651	.873

Chapter VIII

same ancestor through both parents. This probability can be calculated directly from a pedigree by analyzing the paths of genomes descending from the ancestors of an individual (see Fig.100).

With this equation, all possible genetic paths are followed from a designated starting parent to the common ancestor of the path and on to the other parent; n is the number of ancestors in a path including both parents, and F_A is the inbreeding coefficient of the common ancestor of a path. The coefficients for each separate path are summed. This method was developed for calculating inbreeding coefficients in diploid organisms, like sheep, but can still be applied to haplodiploid honey bees as long as the following rules are followed:

1. Identify all common ancestors in the pedigree.
2. Count all females, but never males, in all paths from, and including one parent of the individual being evaluated, to and including the common ancestor of the path, and back to and including the other parent.
3. Paths must go through all ancestors in a path, male and female, even though the males are not counted.
4. If a male is the common ancestor of a path, pass through him and his mother without counting either.
5. Always go against the arrows in the pedigree from the same parent to the common ancestor, and with the arrows returning to the other parent.
6. Ignore the inbreeding coefficients of individuals in the path, except the common ancestor *of that specific path*.

This same method can be used to calculate the *genetic relationship* between individuals. The genetic relationship of two individuals is the proportion of their genes that they share in common because they share ancestors, including parents. Another way to think about a genetic relationship is that it is the probability that two randomly drawn alleles, one from each of two individuals, say X and Y, are identical because they came from the same ancestor. The coefficient of relationship, R_{XY}, can be calculated by using the formula presented in Fig. 100. With this formula n is the number of steps from, but not including, one individual through a common ancestor to the other individual. Do not count male common ancestors. F_X and F_Y are the inbreeding coefficients of the two individuals being evaluated.

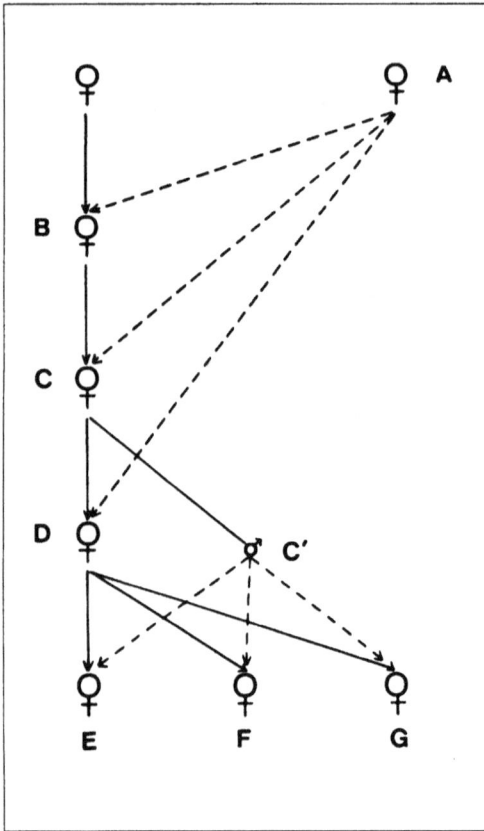

Figure 100. Irregular pedigree.

Figure 100. An irregular pedigree. [From Crow and Roberts,1950. *Genetics* 35: 613-621. Reproduced by permission of the Genetics Society of America.] Solid lines represent egg gametes while broken lines represent sperm.

(1) Calculating an inbreeding coefficient using the formula:

$$F_x = \Sigma (1/2)^n (1 + F_A).$$

n is the number of femal ancestors beginning with one parent of X, through a common ancestor and ending with the other parent. F_A is the inbreeding coefficient of the common ancestor in that path. Common ancestors are underlined. Males are passed through without counting.

Path for queen C:	$B\underline{A} = (1/2)^2 = .25$
Path for queen D:	$CB\underline{A} = (1/2)^3 = .125$
	$C\underline{A} = (1/2)^2 = .25$
	$\Sigma = .375$
Path for queen E:	$D\underline{A}BC = (1/2)^4 = .0625$
	$D\underline{A}C = (1/2)^3 = .125$
	$D\underline{C} = (1/2)^2 (1 + .25) = .3125$
	$\Sigma = .5$

Calculating genetic relationship between two individuals, E and G, using the formula:

$$R_{EG} = \Sigma (1/2)^n \frac{1 + F_A}{\sqrt{(1 + F_E)(1 + F_G)}}$$

n is the number of steps in a path from one individual of the pair through all ancestors to a common ancestor and back to the other individual. The paths from E to G are:

$$\underline{C}'G = (1/2) = .5$$
$$\underline{D}G = (1/2)^2(1.375) = .34375$$
$$D\underline{C}G = (1/2)^3(1.25) = .15625$$
$$D\underline{A}CG = (1/2)^4 = .0625$$
$$D\underline{A}BCG = (1/2)^5 = .03125$$
$$\underline{C}DG = (1/2)^3(1.25) = .15625$$
$$C\underline{A}DG = (1/2)^4 = .0625$$
$$CB\underline{A}DG = (1/2)^5 = .03125$$
$$\Sigma = 1.34373$$

$$R_{EG} = \frac{1.34373}{\sqrt{(1 + .5)(1 + .5)}} = 1.34373/1.5 = .8958$$

Inbreeding Designs

The following are examples of mating designs that may be employed for producing inbred lines for inbred-hybrid breeding programs. These are presented because they are regular systems for which calculating inbreeding coefficients are relatively easy and because they are systems that may actually be useful. The equations shown in Table 4 are recursive. That is, you can use them to calculate the inbreeding coefficient in the current generation (t) based on the past generation coefficients (t-1, t-2, etc.)

• **Self Fertilization.** Queens can be self fertilized by instrumentally inseminating them with sperm from their own drone "sons" (Fig. 101). The sperm gametes of the drones contain genomes derived unchanged from the queen and are equivalent to gametes derived directly from her. To produce inbred lines through self fertilization, virgin queens are given two treatments of carbon dioxide, CO_2, spaced at one or two day intervals. For each treatment, the queen is placed into a container that is flooded with CO_2. The queen is removed from the container as soon as she is unconscious. Don't let her remain in pure CO_2 for more than a few minutes or the treatment may be fatal.

Queens should be provided with drone comb for egg laying; however, young queens often "refuse" to lay in drone sized cells. If this is the case, the queen may be confined to the drone comb under a "push-in" cage to "force" her to lay in drone cells. If all attempts fail to induce her to lay in drone cells, let her lay in worker-sized cells. The nurse bees will still raise the drones, although they will be smaller in size and it will be more difficult to collect their semen for instrumental insemination. Combs containing eggs and hatching larvae may be moved to a queenless nursery colony to insure the best care for the developing drone larvae.

When her drones begin emerging, cage the queen so that her ovaries reduce in size. This will then make it safer and easier to inseminate her. The queen can be inseminated when the adult drones have matured for about 10 days.

• **Mating More than One Queen to a Single Drone.** The semen from a single drone can be divided and used to inseminate more than one queen (Fig. 102). This works best if the sperm are mixed with a diluent to a volume sufficient to inseminate each

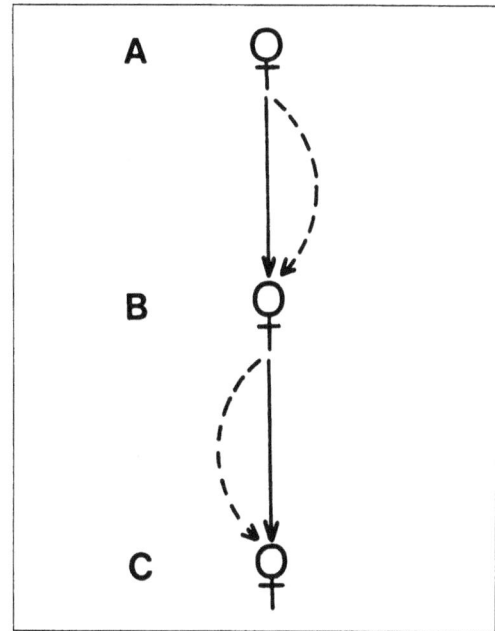

Figure 101. Self fertilization. A virgin queen is treated with carbon dioxide to induce egg laying. A drone is raised from one of her eggs and his semen is used to inseminate the queen.

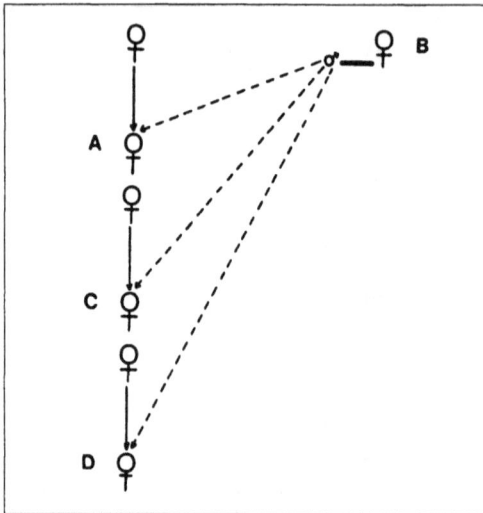

Figure 102. More than one queen mated to a single drone. This can be accomplished by dividing the semen from a single drone and instrumentally inseminating more than one queen.

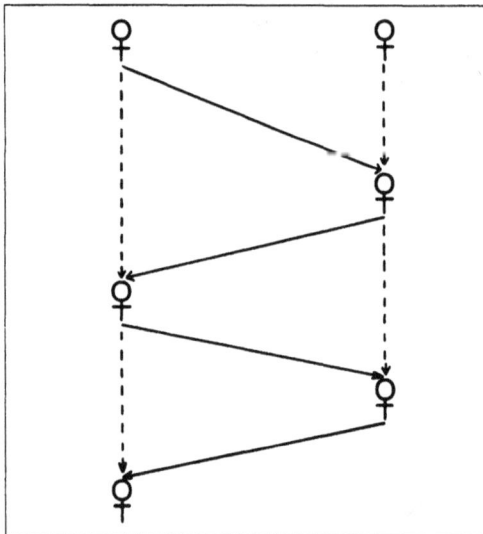

Figure 103. Mother-daughter mating. Virgin queens are produced from a mated queen and are inseminated with semen from their mother's own drones.

queen with about 2ul of total fluid. Several diluents have been reported but a simple one that works is a solution of 0.85% table salt (NaCl) and 0.25% dihydrostreptomycin sulfate (an antibiotic). Sperm should not remain in this solution longer than necessary because it does not contain any nutrients. Other nutrient rich diluents have been developed, but most of the ingredients are difficult for beekeepers to obtain.

• **Mother-Daughter Mating.** Mating a queen to her mother results in the fastest rate of inbreeding because of its relatively short generation times and its relatively high inbreeding coefficient (Fig. 103). Virgin queens are inseminated with semen from drones derived from their mother.

• **Father-Daughter Mating.** Queens can also be mated to their genetic fathers. Drones are raised from the drone mother of the queen and used to inseminate her daughter. This results in repeatedly backcrossing to the same drone mother and results in a maximum inbreeding of 50%.

• **Super Sister-Super Sister Mating.** Super sister matings have greater inbreeding per generation than mother daughter matings but actually proceed more slowly because they have longer generation times (Fig. 104). Drones are raised from a queen and are used to inseminate her super sisters. One advantage of this breeding system is that queens used as drone sources (drone mothers) may mate naturally because their drones are derived directly from their unfertilized eggs. Queen mothers, on the other hand, must be inseminated with the semen of single drones.

• **Full Sister-Full Sister Mating.** Full sister matings are possible but are difficult to conduct because you must be able to determine if sister queens from the same mother queen were derived from the same or different "brother" drones (Fig. 105). Inbreeding proceeds at a per generation rate equal to that of mother-daughter mating, but generation times are greater.

Another way to get full sisters is for two queens to produce both drones and virgin queens. Each is mated to a drone derived from the other. Queens and workers of the different colonies derived from these crosses are full sisters because they share the same mother and father in common. (However, the mother of one is the father of the other.) This mating scheme requires that drones are produced from at least one of the pair of queens before she is inseminated. It is unlikely that this method will be used for commercial breeding but it may be of special interest to researchers.

• **_Half Sister-Half Sister Mating._** Half sisters can either have the same father, (paternal half sisters, Fig. 106), or the same mother, (maternal half sisters, Fig. 107). Half sister matings require three queens each generation.

• **_Cousin Mating._** There are various ways that cousin matings can be applied to produce inbred lines. Inbreeding is relatively slow but eventually lines may become completely homozygous, providing only related individuals are mated to each other.

Many more systems of breeding are possible, for example aunt-niece or grandmother-granddaughter. The kinds of systems possible are limited only by the imagination of the queen breeder.

• **Specific and General Combining Ability.** Specific crosses are performed between lines that are designed to take advantage of either specific or general combining ability of inbred lines. Some lines produce superior performing progeny in combination with many different lines while others are very specific in their combinations. These combining patterns may be explained on the basis of fixation of dominant alleles at loci affecting important traits. For example, assume that there are 4 genetically variable gene loci affecting a trait, say honey production. Each locus has a dominant allele that results in higher honey production and each gene is equal in its effect and all genes are additive. For example AABBccdd results in two "units" of honey production, the same as AaBbccdd. AABBccDD results in three units, the same as AaBbccDd, etc. Now assume that you inbred and crossed 4 lines, W-Z and their genotypes, crosses, and values for honey production are as shown in Fig. 108. Lines W and Z each have values of 2 while lines X and Y have values of 1. However, in combination some crosses have genotypic values higher than either of the parental lines; they show hybrid vigor. Line W shows strong specific combining ability with line Z while line Z shows good general combining ability with all lines.

Advantages and Disadvantages. Inbred-hybrid breeding programs can produce superior uniform stocks. The requirement for a successful program is to produce several lines and test them for

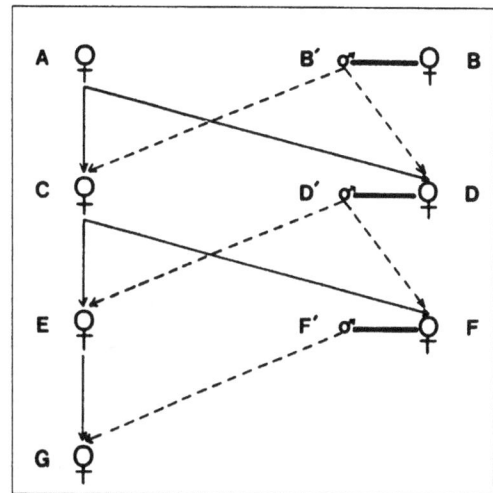

Figure 104. Super-sister mating. One sister produces drones that are used to inseminate the other.

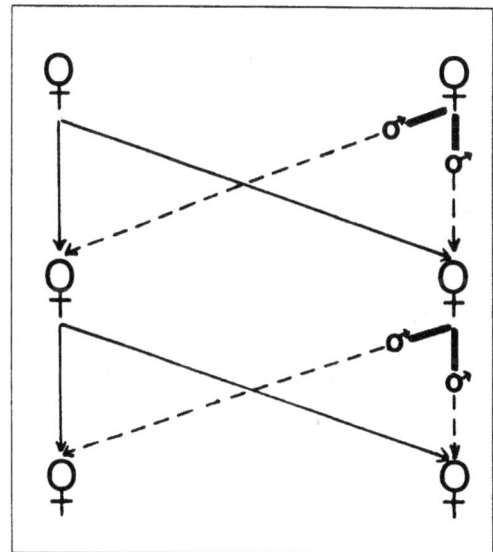

Figure 105. Full-sister mating.

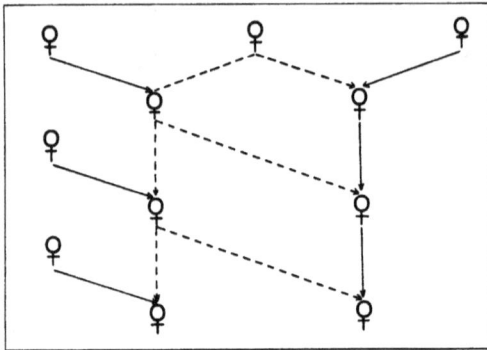

Figure 106. Paternal half-sister mating.

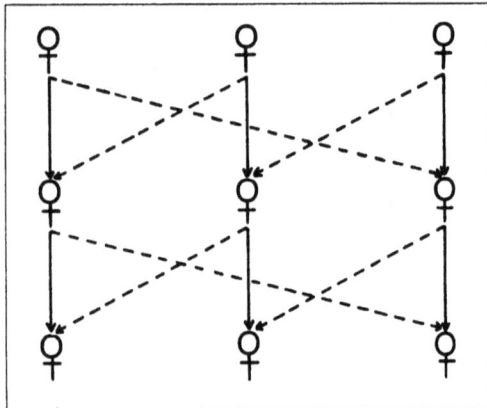

Figure 107. Maternal half-sister mating.

LINE	GENOTYPE	CROSS	VALUE
W	AAbbCCdd	WX	2
X	aabbCCdd	WY	2
Y	AAbbccdd	XY	2
Z	aaBBccDD	ZW	4
		ZX	3
		ZY	3

Figure 108. Hypothetical inbred lines shown with their genotypes, specific hybrid crosses and their hypothetical genotype values in terms of "units" of honey production.

general and specific combining ability. Lines that do not do well in combination with others should be discarded. Some lines will be accidentally lost or will suffer inbreeding depression and will be incapable of surviving. Therefore, it is necessary to continue to produce new inbred lines to replace those that are lost.

Hybrid colonies are very uniform in character. This is because genotypic variability is drastically reduced. For instance, if we cross lines W and Z from Fig. 108, all workers of the resultant colony will have the genotype AaBbCcDd. Therefore, there is no genotypic diversity in this colony, even though every individual has maximum genetic diversity by being heterozygous at all loci.

Colonies containing inbred stocks may be maintained by supplementing them with non-inbred workers. Two methods may be employed: 1) bees or emerging brood may be added periodically to maintain a viable worker population. 2) queens can be inseminated with a combination of sperm from drones of the inbred line and drones from unrelated stocks. The workers derived from the cross of the inbred queen and the unrelated drones will not be inbred and should be able to maintain the colony. However, care must be taken to insure that queens raised from that colony are derived from the appropriate drone father. One way to do this is use distinctly marked colors. For instance, the inbred line is black and the unrelated drone source is yellow.

Care must always be taken when selecting drones for inbreeding. Drones should be marked as they emerge from the comb in order to be sure they are from the desired colony. In addition, colonies should be maintained queenright with unrestricted queens to discourage egg laying by workers. Laying workers produce drone progeny that can result in a mismating and the loss of an inbred line.

Circular Mating Systems

Circular mating systems have been used successfully to produce strains of honey bees that collect and store (hoard) high and low quantities of pollen. Rick Hellmich, a student of Walter Rothenbuhler at the Ohio State University, successfully produced a strain of honey bees that stored much more pollen than another strain that was selected for low pollen storage. A two-way selection program for stored pollen was also conducted at the University of California Davis by Robert Page and Kim Fondrk. After a single generation of selection the high strain colonies contained

Chapter VIII

about twice as much stored pollen and after 3 generations contained more than 5 times as much as low strain colonies. Selection methods are described below.

• **Selection for Pollen Hoarding.** Commercial colonies located in almond orchards near the campus of the University of California were the sources from which the high and low pollen hoarding strains were initially selected. About 400 colonies were examined to estimate the number of frames of bees in each and 127 were selected for further evaluation because they had similar populations of workers, between 6 and 10 frames each. It is important to measure colonies that are nearly equal in worker population and stage of growth and development. The area of stored pollen was then measured in each of the 127 colonies by removing the frames and placing a 1 square inch wire grid over the frame (Fig. 109).

The colonies with the greatest and least quantities of stored pollen were selected from each apiary for a total of 10 high and 10 low pollen hoarding colonies to constitute the high and low strains, respectively. Queens from the high colonies were designated H1 - H10 while those from the low colonies were designated L1-L10. Virgin queens were raised from H1-H5 and L1-L5 and instrumentally inseminated with drones from H5-H10 and L6-L10, respectively (Fig. 110).

Five to ten queens from each subline were inseminated with sperm from single drones. These initial crosses then constituted low strain sublines A-E and high strain sublines Q-U. Sublines were designated on the basis of maternal origin, even though the lines become genetically mixed with additional generations. Following the establishment of the 5 sublines within each strain, vir-

Figure 109. Kim Fondrk making comb measurements using a 1 square inch grid.

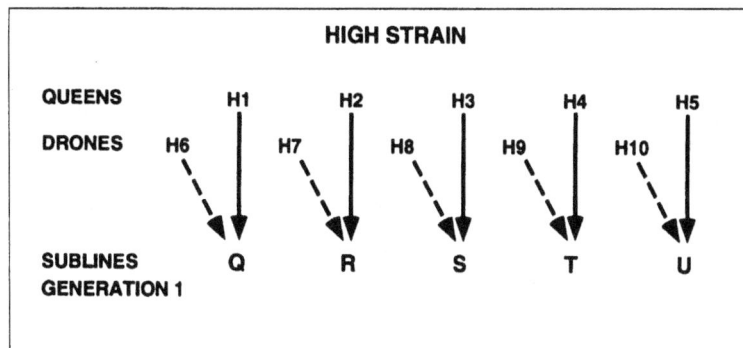

Figure 110. Mating scheme for initial generation of pollen hoarding selection program. From Page and Fondrk 1995. *Behavioral Ecology and Sociobiology* 36: 135-144. Used with permission.

Figure 111. Mating scheme for subsequent generations of pollen hoarding selection program. From Page and Fondrk 1995. *Behavioral Ecology and Sociobiology* 36: 135-144. Used with permission.

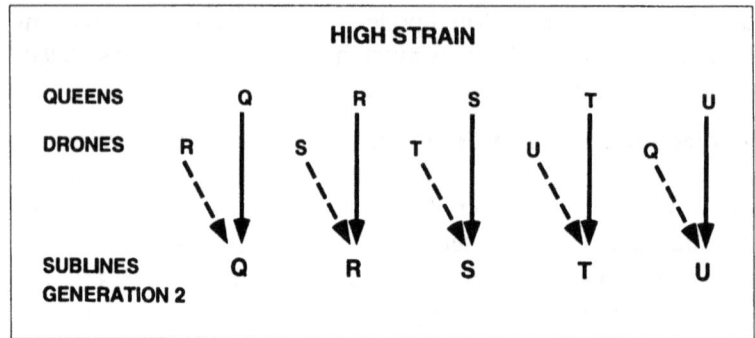

gin queens and drones were raised from the single colony from each subline with the highest or lowest amount of stored pollen (for the high or low strain, respectively). Each generation, 5-10 virgin queens of one subline were instrumentally inseminated with semen from a drone of a different subline, within each strain. A different subline was used in rotation as a drone source for each queen line, each generation (Fig. 111).

• *Advantages and Disadvantages.* The main advantage to a circular mating design is that it delays inbreeding. The effects of inbreeding, such as shot brood occurring from homozygosity at the sex locus, can diminish the ability of a breeder to select the colonies that have superior genotypes for the selected traits. This advantage, however, is only temporary. For instance, in the case of the low pollen hoarding strain colonies, in generation 2 the following crosses were performed: AxB, BxC, CxD, DxE, and ExA. In generation 3, the new queens derived from these crosses receive genomes from two different sublines. The drones for these crosses are still "pure" because they are derived directly from the maternal generation so the next set of crosses are: (AB)xC, (BC)xD, (CD)xE, (DE)xA, and (EA)xB. The fourth generation is (ABC)x(DE), (BCD)x(EA), (CDE)x(AB), (DEA)x(BC), and (EAB)x(CD). In the fifth generation, new queens of all sublines are related and inbreeding begins, for example: (ABCDE)x(EAB) would be the next cross for the A subline.

Another advantage of closed circular mating systems is that selection for a particular trait can be strong and selective progress can be rapid. Selection will progress most rapidly if few sublines are selected from among a large number of commercial colonies and if many new colonies are produced per subline each genera-

Chapter VIII

tion. However, with fewer sublines the response to selection should reach a plateau more quickly, with less overall improvement (Fig. 112). Inbreeding will also become a problem in a relatively few generations which will then require that sublines be outcrossed and rescreened to introduce new genes into the mating population in order to continue the circular mating design. After the first round of crosses, when inbreeding begins to occur, each subline could become an inbred line by using one of the inbreeding mating designs discussed above, then sublines could be tested against each other for combining ability.

Closed Population Breeding and Mass Selection

Difficulties associated with maintaining inbred lines and circular breeding programs stimulated the re-evaluation of *mass selection* methods of breeding honey bees. Mass selection has been used successfully for hundreds or even thousands of years by agriculturally based societies to produce superior staple food crops like corn and wheat. For commercial honey bees, mass selection is performed when a large number of colonies are produced from a selected set of parents (queens of superior performing colonies), and are evaluated and selected as parents for the next generation on the basis of their colony performance. A population is *closed* when matings are controlled so that only offspring of selected parents are used as parents for the next generation. The result is the mating population is closed to any genetic material from outside of the breeding program. Inbred-hybrid and circular mating programs are both closed population programs.

Closed population mating with mass selection was originally professed by Professor Harry Laidlaw as a needed change in direction for honey bee breeding, then was developed into a formal breeding theory by both authors of this book (Laidlaw and Page). The implementation of breeding programs based on their theoretical analyses soon followed in the United States, Canada, Australia, the former Soviet Union, Egypt, and several European countries, with many claims of success. Perhaps the most successful commercial program was the production of the "New World Carniolan" stock by Sue Cobey and Tim Lawrence, and is discussed below.

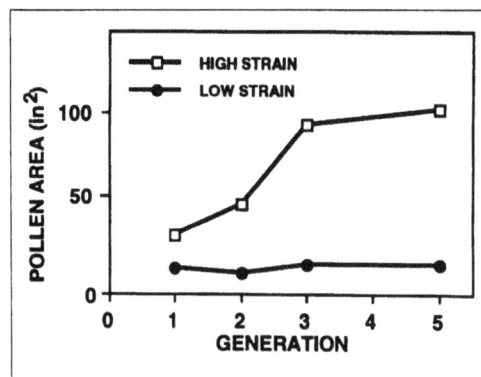

Figure 112. Results of five generations of selection for high and low quantities of stored pollen.

• **Breeding Theory.** One of the main objectives of the closed population breeding program developed by Page and Laidlaw is to maintain as many sex alleles as possible in the breeding population. Many sex alleles are needed to reduce the production of diploid males that results in reduced brood viability, a condition known as "shot brood". Sex alleles are lost from a population as a consequence of accidents of sampling. For example, imagine that you have a bucket full of marbles of 10 different colors. Each color is equally represented among the marbles. What chance would you have to draw 10 marbles from this bucket and get one of each color? Your chance would be quite slim. That is equivalent to sampling 10 drones at random from a population with 10 equally frequent sex alleles. If you sample 5 queens, you also get 10 sex alleles. But in this case you do a little better than when you sample 10 marbles from the bucket, or 10 random drones. Each queen has two sets of chromosomes, and, therefore, two sex alleles. However, each queen has two different alleles, otherwise she would have been eaten as a young larva. So for each queen you select you are guaranteed two different sex alleles. That would be equivalent to drawing 10 marbles from the bucket, two at a time and discarding pairs of marbles that are the same color. Therefore, the number of sex alleles that can be maintained in a population will depend on how many queens are selected each generation as breeders. Each breeder has two sex alleles plus alleles stored in her spermatheca from the sperm of her mates.

Mutation is the source of new sex alleles while sampling accidents are the way sex alleles are lost. Selection against diploid males resulting from homozygosity at the sex locus stabilizes sex allele frequencies by favoring rare sex alleles. Rare sex alleles are favored because they are less likely to be homozygous in individuals. For instance, assume that the bucket of marbles contains mostly red marbles. Each time you draw two marbles you have a good chance of getting two that are red. If they were sex alleles, those individuals with two "red" sex alleles would be eaten by the workers, could not become queens, and would result in a decrease in the abundance of that common sex allele. On the other hand, if say blue ones were very rare, every time you drew a blue marble, it would probably be paired with another that was not blue. If blue represented a rare sex allele, it would almost always be viable, would be over-represented in queens relative to the "red" allele, and would increase in frequency from one generation to the next. The "red" allele would decrease. If a population is very large, it is expected that all sex alleles will be close to equal in frequency.

For any given sized population, the number of sex alleles should become stable and depends on the rate of mutation and the size of the breeding population. For breeding programs, the rate of mutation of sex alleles into new alleles is too low to have any significant effects on increasing the number of alleles. Populations of honey bees have been estimated to contain between about 6 and 17 different sex alleles. That requires hundreds or thousands of breeder queens in order to be stable over time, impossible for commercial breeding programs. So, the main concern for honey bee breeders, is how many generations a selective program can operate before it demonstrates an economically damaging reduction in brood viability.

Computer simulations have provided some answers to this question. For these simulations, it was assumed that a minimum brood viability of 85% was needed for commercial colonies and that the expected "life" of a breeding program is no more than about 20 years with one generation per year. It was also assumed that the commercial populations from which breeder colonies are extracted for breeding programs initially contain 10 equally frequent sex alleles. For example each sex allele is represented at 10%. Each computer "generation", queens and drones were produced from each queen and each virgin queen was mated to 10 randomly selected males. New breeder queens for producing the next generation were then selected from the pool of mated daughter queens. New breeders were selected in two ways: 1) randomly, without regard for maternal source, or 2) each breeder queen was replaced by one of her randomly chosen daughters. These are called the *random selection* and *queen supersedure* models, respectively.

Computer simulations demonstrated the following: 1) 35-50 breeders are needed with random queen selection to maintain at least 85% brood viability for at least 20 generations. 2) The number of breeders may be reduced to 25 if breeder queens are replaced with their own daughters, queen supersedure. Other simulation studies have shown that sex alleles are lost more slowly if brood viability is used as one of the selection criteria, suggesting that even fewer breeders may be used in closed populations with little loss of sex alleles and brood viability.

• **New World Carniolan Program.** Sue Cobey and Tim Lawrence established their initial breeding population by collecting queens from the United States and Canada that presumably descended from Carniolan stock. Potential breeder queens were first tested to insure that their progeny had characteristics of Carniolans before they were included into the closed breeding population. Following the "random selection model" of Page and Laidlaw, the breeding population consisted of 35-50 breeder queens from which both virgin queens and drones were raised. Each generation, 5-10 daughter queens were instrumentally inseminated with semen that was collected from an equal number of drones derived from each breeder. Semen from different males was pooled, homogenized, and collected into a large volume syringe for inseminating several queens from each batch. As a consequence of pooling and homogenizing the semen, each queen was effectively mated to the same batch of drones that represented the *gene pool* of the entire closed population. Differences in colony performance among queens was due to differences in the genotypes of the queens, rather than the drones with which they mated. This greatly facilitated the selection of queens with superior genotypes.

Each year a new generation of 175-250 instrumentally inseminated queens was produced. Queens were introduced into colonies that grew in population and were used as commercial production hives. Colonies were evaluated to select the 35-50 queens to replace the previous generation as breeders. Evaluations were performed in two steps. For the first step, colonies were preselected on the basis of brood viability, temperament (defensive behavior), spring buildup (worker population), cleaning behavior (removal of debris from hive), occurrence of disease, swarming (presence of swarm cells), and color (black). Colonies that were not eliminated from the pool during the prescreening were then evaluated for honey production.

Honey production was estimated by using a technique developed by Dr. Tibor Szabo for a closed population breeding program he developed and directed for Agriculture Canada. Szabo's program successfully produced superior honey producers by evaluating short-term weight gains. During a nectar flow, colonies were supered and then weighed to establish a base weight. Ten days later, all colonies were again weighed and the original weight subtracted. The difference between the two weights represented the short term gain in honey production and was used as the final selection criterion. Those 35-50 colonies with the highest weight gain became the breeders.

• ***Advantages and Disadvantages.*** Closed population breeding with mass selection offer opportunities to selectively improve honey bee stocks beyond what can be achieved by inbred-hybrid breeding. With inbred-hybrid breeding the best stock possible is no better than the best possible in the original population. By adding selection, it is possible to recombine genotypes into novel combinations that did not exist in the original population and produce colonies with phenotypes that are outside the original range.

The "random selection" scheme has the advantage that all colonies are selected on the basis of their performance, not their pedigree. Therefore, selection intensity is stronger and the response to selection should be faster. Queen supersedure, however, has the advantage that fewer colonies may be maintained as breeders, but selection is expected to be somewhat slower. Uniformity of stock will take longer to achieve with both methods than is possible with inbred-hybrid breeding.

A Simple Program for Beekeepers

Most beekeepers don't have the resources to conduct an extensive breeding program such as those outlined above. However, they can improve their stocks if they eliminate queens in colonies that have bad characteristics, such as brood diseases, strong defensive behavior, poor productivity, etc., and replace them with queens derived from good colonies. In addition, all colonies should be requeened regularly with good stock in order to provide a pool of drones of good quality to mate with new queens produced throughout the season. In time, the general quality of all colonies should improve, however, not to the degree that may be obtained with a more vigorous program.

Chapter IX

THE GENETIC BASIS OF DISEASE RESISTANCE

Beekeeping problems resulting from diseases and parasites are continually becoming more abundant, costly, and difficult to control. Typically, the solutions have been chemical applications to hives with the inherent risks of contaminating wax and honey, and developing pesticide resistance in the target organism. The use of chemical remedies is also becoming less acceptable to the public, leading to increasingly more restrictions in their use, particularly around food products like honey. Therefore, in the future breeding resistant stocks of honey bees will become increasingly important for maintaining a viable beekeeping industry. As a consequence, this chapter on selecting for disease resistance has been added with the hope that it will stimulate an interest in breeding for resistance and guide those queen breeders who so endeavor.

Mechanisms of Resistance

The basis of all resistance to diseases and parasites is some mechanism whereby the host, the honey bee, defeats the disease agent. In order to select for resistance, there must be genetic variability for a mechanism. Genetic variability has been demonstrated for three general classes of disease resistance in honey bees: *physiological*, *behavioral*, and *anatomical*. Mechanisms of more than one type may simultaneously operate against the same pathogen or parasite. Also, the same mechanism may operate against more than one pathogen or parasite.

Physiological Mechanisms
Physiological resistance is what first comes to mind when we think of resistance to diseases. Somehow, the larva or adult bee produces some physiological product that slows or inhibits the growth, development, or reproduction of the disease agent. Genetic vari-

ability for this kind of resistance has been demonstrated for American foulbrood (AFB) disease by Professor Walter Rothenbuhler of the Ohio State University, and his associates. A more rapid rate of larval development is another possible physiological mechanism of larval resistance to AFB. Larvae that develop faster may be less likely to die from AFB infection. In fact, larvae of an AFB resistant strain of Rothenbuhler's (the "Brown line") were larger than larvae of his susceptible strain during early larval development, perhaps demonstrating this kind of resistance mechanism. Nurse bees derived from stocks imported from Hawaii, that were believed to be resistant to AFB, conferred resistance to the larvae they fed when colonies were inoculated with AFB spores. The germination of spores cultured in brood food derived from the resistant Brown line was inhibited.

Another kind of physiological resistance is shown by honey bee larvae to varroa infestation. Colonies with workers that develop faster into adults have shorter periods of time after they are capped and are relatively resistant to varroa compared with colonies with longer post-capping stages. This resistance is a consequence of the necessity for varroa to develop on larvae within a capped cell. Fewer mites develop in cells containing workers with shorter post capping periods because there is insufficient time for the mites to develop from eggs into adults.

Behavioral Mechanisms

The best known example of disease resistance in honey bees is the hygienic behavior which was originally described by Professor O.W. Park in 1937. Apparently, there are two independent behavioral activities: uncapping cells and removing diseased larvae from uncapped cells. These two activities were shown by Rothenbuhler in 1964 to be under the control of two independent genetic mechanisms. Hygienic behavior has also been shown to be an effective behavioral mechanism against chalkbrood disease and infestations of varroa.

Recently, there has been considerable interest in the effects of grooming behavior of bees on colony levels of varroa infestation. This interest is a consequence of the finding that workers of the Asian honey bee, *Apis cerana*, groom mites from their bodies and kill the mites by puncturing them with their mandibles.

Anatomical Mechanisms

Honey bees differ in their abilities to reduce the numbers of AFB spores in stored honey. It is believed that the *proventricular valve*, also called the *honey stopper*, that is located in front of the *crop*, or *honey stomach*, filters the spores from the nectar or honey consumed by workers. Workers of the AFB resistant Brown line of Rothenbuhler were shown to be more efficient than susceptible bees at removing spores from sugar syrup containing suspensions of spores of *Bacillus thuringiensis*, *Bacillus cereus*, and the AFB causative agent, *Bacillus larvae*.

Selection Programs

There are remarkably few documented examples of controlled breeding for resistance to honey bee diseases and parasites. This is unexpected when considering the economic importance of honey bees, and the dramatic results obtained in those few cases where selective breeding was practiced. Here, we discuss some of the successful breeding programs.

Resistance to American Foulbrood

The first successful breeding program for resistance to AFB was implemented in September, 1934, by O.W. Park, F.B. Paddock, and Frank C. Pellett. Their breeding program was a more-or-less closed population program with more-or-less controlled isolated mating, and mass selection. Over the next 15 years they successfully selected a resistant stock and documented their progress (Fig. 113).

• *Source of Bees.* They initiated their breeding program by requesting beekeepers to supply them with colonies that they thought were resistant to AFB. Altogether they received 45 colonies from throughout the United States representing Carniolan, Italian, and Caucasian races. Colonies were consolidated in a single apiary near Atlantic, Iowa, and were tested for resistance.

• *Assay.* Each colony received one rectangle of brood comb containing approximately 200 cells with about 75-100 AFB scales. The comb sections were inserted into brood combs in the center of the brood nest. Colonies were of equivalent strength, occupying at least 8 Modified Dadant brood combs. Colonies were frequently examined for evidence of AFB disease.

• *Mating System.* Presumably isolated mating yards were established. Queens and drones of resistant stocks mated naturally from these apiaries, however, it is unknown how much outcrossing actually occurred due to incomplete isolation.

• *Mechanisms of Resistance.* Park demonstrated that one mechanism of resistance was behavioral. He found that some colonies quickly tore down and removed comb inserts containing scales of AFB. Those that removed the diseased combs had a lower incidence of disease. He also demonstrated that they were responding to the presence of the disease, not just the foreign comb. Park, however, suspected that there were other, physiological, mechanisms involved, besides house cleaning behavior.

Rothenbuhler, after reviewing the work of Park, also believed that hygienic behavior alone was not sufficient to explain all resistance to AFB. As a consequence, he initiated a breeding program in 1954 specifically to study the potential genetic mechanisms of resistance.

• *Source of Bees.* In 1954, Rothenbuhler obtained ten queens from Mr. Edward G. Brown that were believed to be resistant to AFB. Mr. Brown had operated a wax rendering plant for many years. During this time he had been selecting for disease resistance in his apiary by leaving diseased combs, brought to him for rendering, exposed to robbing bees. Queens from this apiary constituted the foundation for the resistant "Brown" line.

Rothenbuhler also started a susceptible line. He believed that the best way to study resistance was to compare putative resistant stocks to those that are susceptible. Therefore, in 1950 he obtained a single, mated queen from Mr. Homer Van Scoy from which he initiated a disease-susceptible line, the "Van Scoy" line.

• *Assay.* The assay used by Rothenbuhler to breed for resistance and susceptibility was the same as that of Park discussed above.

• *Mating System.* Mr. Brown raised new queens from his contaminated apiary and allowed them to mate naturally resulting in a kind of "natural" selection for disease resistance. Professor Rothenbuhler used instrumental insemination to maintain inbred lines of his resistant and susceptible stocks.

• *Mechanisms of Resistance.* Using these resistant and susceptible strains, Rothenbuhler and his associates demonstrated one

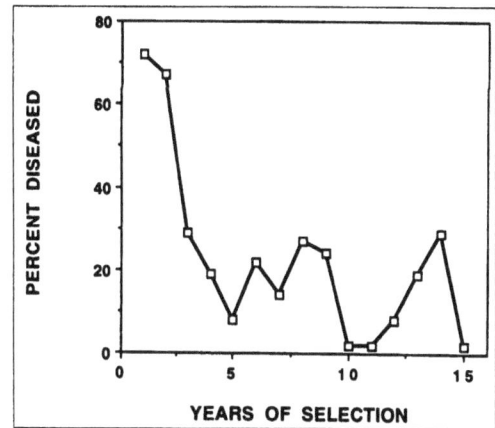

Figure 113. The percentage of AFB innoculated colonies that became diseased during 14 years of selective breeding. Data are from the selection porgram of O. W. Park covering the years 1935-1942 as presented in Table II, Rothenbuler 1958. *Annual Review of Entomology* 3: 161-180.

behavioral, three physiological, and a single, putative anatomical mechanism (see discussion of mechanisms above). The results of studies of AFB resistance demonstrate the diversity of resistance mechanisms that can simultaneously occur when colony selection is used. Colony selection focuses on the occurrence of the disease, not the specific mechanisms responsible for resistance. Single factor resistance may occur if only a single mechanism is selected, for example hygienic behavior. However, multifactor resistance is probably more effective.

Resistance to Tracheal Mites

Several studies have revealed significant differences in resistance among North American commercial stocks. Professors Robert Page and Norman Gary conducted two-way selection for the susceptibility of young workers to infestation by adult female mites. They used instrumental insemination and a circular mating scheme. After a single generation of selection, workers from the susceptible strain were 40% more likely to become infested than workers of equal age from the resistant strain. After the second generation of selection, the susceptible strain workers were 2.4 times more likely to become infested. They also tested their generation 2 high and low strain bees against local commercial bees. The commercial bees were intermediate between the selected strains for levels of tracheal mite infestation.

• *Source of Bees.* The breeding population used by Page and Gary was derived from a test of phenotypic variability conducted on 22 colonies derived from queens of a closed honey bee population developed and maintained in Madison, Wisconsin. An additional queen from a commercial source was included in their study for a total of 23. The closed population was originally established from 25 queens obtained from queen producers located throughout the United States. Colonies were tested for worker susceptibility and those queens that produced worker progeny with the highest and lowest likelihood of becoming infested were selected as queen and drone mothers for the susceptible and resistant lines, respectively.

• *Assay.* Page and Gary used two types of assays for their studies. The first was a laboratory assay where they caged combs of emerging brood and placed them in an incubator. Workers from all colonies to be tested, the "source colonies", emerged during the same 24 h period and were individually tagged with plastic numbered tags for identification, then placed into screen-wire cages. Each of several test cages contained 10 workers from each source colony, the "target" bees, plus an equivalent number of workers taken from the top bars of the brood nest of a highly infested colony, the "host" bees. Target bees remained in cages in the incubator, in contact with host bees for 7 days, after which time all target bees were sacrificed, dissected, and the number of mites counted in each prothoracic tracheal trunk.

They concurrently tested target bees in highly infested host colonies. Ten workers from each source colony were tagged in the same way as for the cages, then placed in host colonies for 5-7 days, after which they were removed and dissected. All breeder queens, each generation, were selected on the basis of the laboratory test. The field colony assays were used to confirm that selection based on the laboratory assay was resulting in differences in susceptibility among workers in field colonies.

• *Mating System.* Two way selection was performed. Relatively resistant and susceptible breeding populations were established from the original foundation population of 23 that was tested for phenotypic variability. Queens from the foundation population were ranked from 1 to 23 on the basis of the average number of adult mites found in tracheae of their progeny in the cage studies. Each generation, drones and several virgin queens were raised from surviving queens with high (most mites per worker) and low ranks (fewest mites per worker) to produce the next generation of susceptible and resistant workers, respectively. For generation 1, 4 resistant and 3 susceptible queens were selected as breeders, for generation 2, 5 resistant and 4 susceptible queens were used. Each virgin queen was instrumentally inseminated with the semen from a single drone that came from a corresponding, unrelated high or low ranked queen.

• *Mechanism of Resistance.* The mechanism of resistance is unknown.

Resistance to Varroa

Drs. Jovan Kúlincevic and Tom Rinderer initiated a two-way breeding program with A. *m. carnica* colonies in Yugoslavia. The pa-

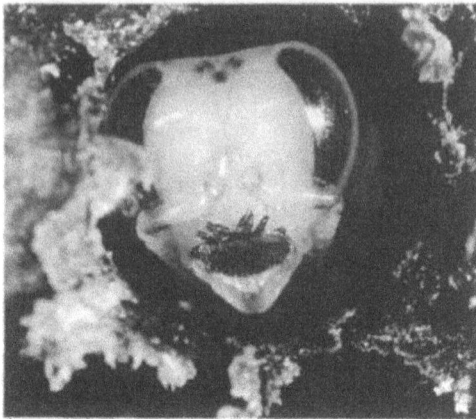

Figure 114. Varroa female on the face of a worker pupa.

rental generation of the susceptible line averaged 14.8% of brood cells infested while the parental colonies of the resistant line averaged 5.6%. After four generations of bidirectional selective breeding, resistant colonies had about half the number of infested cells as the susceptible colonies. Queens from the resistant strain were imported into the United States and were tested against unselected U.S. stocks. Unfortunately, the Yugoslavian bees were no more resistant than any of the U.S. stocks against which they were compared.

• *Source of Bees.* The breeding population was derived from *A. m. carnica* colonies that had survived a varroa infestation in Yugoslavia.

• *Assay.* The evaluation assay was to estimate the proportion of capped worker brood cells infested by reproducing mites. In each colony, 100 capped worker cells were opened and examined to determine the number of female, male, and immature mites. The two queens of the colonies having the highest infestation level were chosen for developing the susceptible strain, whereas the two queens from the colonies with the lowest infestation level, were chosen to develop the resistant strain.

• *Mating System.* Queens were reared from the selected queen mothers, and were permitted to open mate in the same apiary.

• *Mechanism of Resistance.* The mechanisms of resistance were not identified.

Potential Mechanisms of Resistance to Varroa

Colony infestation with varroa is the most serious problem for beekeeping, worldwide (Fig. 114). As a consequence, it is remarkable that there are no examples of successful artificial selection for resistance. However, much is known about the biology of varroa and how it interacts with its natural host, *A.cerana*, and its new, accidental host, *A. mellifera*. This understanding of biology suggests potential mechanisms of resistance, some of which have been demonstrated to vary genetically.

Natural Variation in Resistance to Varroa
The western honey bee, *A. mellifera*, is not the natural host of varroa. Its natural host, *A. cerana,* demonstrates several resistance

mechanisms that apparently have evolved in response to a long term association with varroa (see below). However, most temperate colonies of *A. mellifera* succumb to varroa and seem to have little resistance. In spite of this, some colonies survive even in locations that have been devastated by varroa, suggesting that those colonies may have some degree of resistance to the parasite. Although varroa has been a problem for honey bees in most countries, in the South American tropics it does not appear to be a serious pest. The tropical climate and tropically-adapted bees, such as the Africanized honey bees, both play a role in maintaining reduced levels of infestation.

Mechanisms of Resistance

Several mechanisms of natural resistance against varroa have been reported for *A. cerana*. These same mechanisms have been reported to occur at much lower levels in populations of *A. mellifera*, but offer hope for selection programs designed to enhance them.

• *Grooming Behavior.* *A. cerana* has at least four behavioral mechanisms that may contribute to their relative resistance to varroa. Professor Christine Peng, University of California Davis, showed that workers infested with varroa extensively groom themselves to remove the mites. If they cannot remove them, they perform a grooming dance and nearby workers use their mandibles to remove the mites. After removing the mites from a nestmate, a grooming worker frequently punctures the mite with her mandibles and removes it from the nest. When mites are particularly difficult to dislodge, the bee attracts a large number of workers that then engage in group grooming of the infested individual. Evidence for all of these grooming behavioral responses to varroa have been directly or indirectly observed in *A. mellifera*, but at much lower frequencies than occur in *A. cerana*.

Africanized honey bees of South America seem to be better able to defend themselves from varroa than are Italian honey bees. Studies by Geraldo Moretto and his associates have shown that Africanized workers in Brazil are seven times more efficient than Italian workers in eliminating mites from their bodies. In their study, 38.5% of the Africanized workers infested were able to discard the mites from their bodies, as compared with only 5.75% of the Italian workers. This study suggests that selection for this trait is possible.

• *Hygienic Behavior*. A. cerana workers are able to detect capped brood cells that are infested with mites. They open the cells, remove, and kill the mites. Workers of A. mellifera apparently can also detect and remove varroa mites from infested brood cells in a way similar to A. cerana. If an infested cell is uncapped, the immature mites die, and the adult females must search for another cell, or are killed by the bees. This should slow the population growth of varroa in the colony.

• *Duration of the Post-Capping Stage*. The post-capping stage is one of the best studied mechanisms of resistance of A. mellifera against varroa. The amount of time that a bee remains within a capped cell could influence the reproductive rate of varroa. A female mite enters a cell shortly before it is capped then waits about 60 h before laying her first egg which requires about 6.2 - 7.5 days to develop into an adult. Eggs are laid singly at approximately 30 h intervals. Therefore, mites from eggs that were laid with fewer than seven capped-cell days left are not likely to survive. So, if the post capping stage could be shortened, it may be possible to reduce the reproductive output of female mites. The duration of the post capping stage varies with bee race. European honey bees require about 12 days from sealing of the cell to adult emergence, while the Cape bee, A. m. capensis needs only 9-10 days. Africanized bees have a post capping stage of about 11 days. This shorter post capping time may at least partially explain their relative resistance to varroa.

• *The Attractiveness of Brood and Adult Bees.* Varroa females demonstrate a strong bias favoring drone rather than worker cells. If this bias is based on the relative attractiveness of drone and worker brood, then perhaps the reproductive capacity of varroa may be reduced by selecting for less attractive worker brood. Dr. Ernesto Guzmán-Novoa found that European brood is twice as susceptible to varroa infestation as brood from Africanized honey bee colonies. Adult European workers are also more attractive to varroa than are Africanized bees. The reason for this lower attractiveness is not clear, but it may involve qualitative and/or quantitative differences in attractive chemicals produced by the brood and adult workers.

• **Mite Fertility-Limiting Factors.** The proportion of varroa that reproduces varies with host species, host race, and the sex of the brood. Varroa females apparently do not reproduce when they infest worker brood of *A. cerana*. Likewise, varying proportions of female mites do not reproduce on brood of different races of *A. mellifera*. Values for the percentage of female varroa that reproduce on European bees have been estimated to range from 43-84% on worker brood and 86-95% on drone brood, depending on the race of bees studied. If variation for this characteristic is due to honey bee genetic differences, then selective breeding for this trait might lead to the development of varroa resistant stocks.

Perspectives and Conclusion

The world-wide eradication of any honey bee disease is unrealistic, therefore, in order to reduce economic damage the beekeeping industry must depend on methods that maintain pathogens and parasites at reduced levels. The widespread use of chemical treatments has serious potential costs and risks resulting from the evolution of chemically resistant strains of disease agents, and the chemical contamination of hive products. Thus, the development of control methods that do not depend on chemicals should receive more attention. The evidence presented above suggests that there is sufficient genetic variability for resistance to diseases to make selective breeding a viable component of commercial honey bee management.

However, selective breeding programs will not succeed without economic incentives. Current prices paid for queens produced in the United States will not support the added expense of industry-driven breeding programs. Institutional breeding programs at state supported universities, and the United States Department of Agriculture, have never succeeded due, at least in part, to the failure of the bee industry to adopt the stocks they produced. If the beekeeping industry is not willing to pay higher prices for selected stocks and/or is not willing to support and accept stocks produced by institutional stock improvement programs, then the only alternative for the future is the continued use of dangerous and expensive chemicals.

Bibliography

Abbott, C.P. 1947. *Queen Breeding for Amateurs.* King and Hutchings, Ltd., Uxbridge and Southhall, Great Britain.

Adam, A. 1913. Bau und Machanismus des Receptaculum seminis bei den Bienen, Wespen und Ameisen. *Zool. Jahrb. Abt. 2 Anat. u. Ont.* 35:1-74.

Adams, J., E. Rothmann, W.E. Kerr, and Z.L. Paulino. 1977. Estimation of sex alleles and queen matings from diploid male frequencies in a population of *Apis mellifera. Genetics* 86:583-596.

Alley, H. 1873. Sending queens by mail. *Amer. Bee J.* 9:109-110.

Alley, H. 1883. *The Beekeepers Handy Book* or twenty-two years experiences in queen rearing. Publ. by author. Wenham, Mass.

Alphonsus, E.C. 1931. The life of Anton Jansha. *Amer. Bee J.* 71:508-509.

Anderson, J. 1918. Laying workers which produce female offspring. *Amer. Bee J.* 58:192.

Anonymous. 1947. Roberts and Mackensen apparatus for artificial insemination of queen bees. *Amer. Bee J.* 87:425.

Bailey, L. 1955. The epidemiology and control of Nosema disease of the honey-bee. *Ann. Appl. Biol.* 43:379-389.

Beetsma. J. 1979. The process of queen-worker differentiation in the honeybee. *Bee Wld.* 60:24-39.

Berlepsch, The Baron of. 1861. The Dzierzon Theory. Transl. of Berlepsch's "Apistical Letters" by Samuel Wagner, *Amer. Bee J.* 1:1-10 and following numbers of the volume.

Bertholf, L.M. 1925. The moults of the honeybee. *J. Econ. Entomol.* 18:380-384.

Bevan, E. 1838. *The Honeybee. Its Natural History, Physiology, and Management.* Baldwin, Cradock, Jay, London.

Bishop, G.H. 1920. Fertilization of the honeybee. I. Disposal of the sexual fluids in the organs of the female. *J .Expt. Zool.* 31:225-265, 267-286.

Blum, M.S. 1992. Honey bee pheromones. In *The Hive and the Honey Bee,* J.Graham, ed., Dadant and Sons, Inc., Hamilton, Illinois

Boyd, W.L. 1878. Queen cells to order. *Glean. Bee Cult.* 6:323.

Brooks, J.M. 1880. How to get plenty of choice queen cells another way. *Glean. Bee Cult.* 8:362.

Brunnich, Dr. 1913. Fertilizing queens at a mating station. *Glean. Bee Cult.* 41:493-497.

Butler, C. 1634. *The Feminine Monarchy, or the History of Bees.*

Butler, C.B. 1954. The method and importance of the recognition by a colony of honeybees (*A. mellifera*) of the presence of its queen. *Trans. Roy. Entomol. Soc. London.* 105:pt.2.

Butler, C.B. and J. Simpson. 1958. The source of the queen substance of the honeybee (*Apis mellifera*). *Proc. Roy. Entomol. Soc Lond.* 33:120-122.

Cale, G.H. 1926. The first successful attempt to control the mating of queen bees. *Amer. Bee J.* 66:533-534.

Callow, R.W. and N.C. Johnston. 1960. The chemical constitution and synthesis of queen substance of honeybees (*Apis mellifera*). *Bee Wld.* 41:152.

Cobey, S. and T. Lawrence. 1988. A successful application of the Page/Laidlaw breeding program. *Glean. Bee Cult.* 116:274-276.

Columella, L.J.M. 1745. *Of Husbandry.* Printed in 12 books for A. Millar. Book 9:396.

Cook, V. 1986. *Queen rearing Simplified.* British Bee Publications Ltd. Geddington, Northamptonshire, Great Britain.

Dadant, C. P. 1906. *Langstroth on the Hive and the Honey Bee.* Revised 1913. Dadant and Sons, Hamilton, Ill.

Davis, J.L. 1874. Davis transposition process. *Glean. Bee Cult.* 2:107.

Doolittle, G.M. 1915. *Scientific Queen-Rearing.* 6th ed. American Bee Journal, Hamilton, Illinois.

DuPraw, E.J. 1961a. Personal communication.

DuPraw, E.J. 1961b. A unique hatching process in the honeybee. *Trans. Amer. Micro Soc.* 80:185-191.

Dustmann, J.H. 1991. Gen-Wirkungen bei Mutanten der Honigbiene *Apis mellifica:* Zur Biochemie und Histologie der Ommochrom-und Pteridinpigmente. Apimondia-Verlag, Bukarest.

Dustmann, J.H. 1994. Personal communication

Dustmann, J.H., M. Kühnert, P. Schley, and F.K. Tiesler. 1991. *Instrumental Insemination of Queen Bees.* C 1746. Institut für den Wissenschaftlichen Film. Göttingen, Germany. (VCR tape available).

Eckert, J.E. 1934. Studies in the number of ovarioles in queen honey-bees in relation to body size. *J. Econ. Ent.* 27:629-635.

Eckert, J.E. and F.R. Shaw. 1960. *Beekeeping.* Macmillan Company, New York.

Erickson, E.H. and E.W. Herbert. 1980 Soybean products replace expeller-processed soyflour for pollen supplements and substitutes. *Amer. Bee J.* 120:122-126.

Free, J.B. 1960. The distribution of bees in a honey-bee (*Apis mellifera* L.) colony. *Proc. Roy. Entomol. Soc. Lond.* (A):141-144.

Gary, N.E. 1960. Mandibular gland extirpation in living queen and worker honeybees. *Bee Wld.* 41:229.

Gary, N.E. 1961. Queen honeybee attractiveness as related to mandibular gland secretion. *Science.* 133:1479-1480.

Gary, N.E and R.E. Page.1987. Phenotypic variability in susceptibility of honey bees, *Apis mellifera* L.,to infestation by tracheal mites, *Acarapis woodi* Rennie. *Exp. Appl. Acarol.* 3:291-305.

Good, J.R. 1881. Queen cage candy. *Glean. Bee Cult.* 9:374.

Graham, J.M. ed. 1992. *The Hive and the Honey Bee*. Dadant and Sons, Hamilton, Ill.

Haeckel, E. 1896. *The Evolution of Man*. Vol. 1:28. D. Appleton & Co. New York.

Hallamshire Beekeeper (John Hewitt). 1892. Fertile workers—their utility. *Jour. Hort.* London 25:134.

Harbo, J.R. 1979. The rate of depletion of spermatozoa in the queen honeybee spermatheca. *J. Apic. Res.* 18:204-207.

Harbo, J.R. 1985. Instrumental insemination of queen bees. *Amer. Bee J.* 125:197-202, 282-287.

Haydak, M.H. 1943. Larval food and development of castes in the honeybee. *J. Econ. Entomol.* 33:772-792.

Haydak, M.H. and A.E. Vivino. 1950. The changes in the thiamine, riboflavin, niacin, and pantothenic acid content in the food of female honeybees during growth, with a note on the vitamin activity of royal jelly and bee bread. *Ann. Entomol. Soc. Amer.* 43:361-367, Tables 1 & 2.

Heberle, J.A. 1913. Mating stations. *Glean. Bee Cult.* 41:497-498.

Hellmich, R.L., J.M. Kulencevic and W.C. Rothenbuhler. 1985. Selection for high and low pollen-hoarding honey bees. *J. Hered.* 76:155-158.

Herbert, E.W. 1992. Honey bee nutrition. In *The Hive and the Honey Bee*, Joe M. Graham, ed., Dadant and Sons, Hamilton, Illinois. pp 197-233.

Hitchcock, J.D. 1956. Honeybee queens whose eggs all fail to hatch. *J. Econ. Entomol.* 49:11-14.

Holm, E. 1986. *Artificial insemination of the queen bee.* Eigil Holm, Byskovsvej 4, DK-8751, Denmark.

Huber, F. 1814. *New Observations upon Bees.* (Transl. by C.P. Dadant) *American Bee Journal*, Hamilton, Illinois (1926).

Hunt, G.J., R.E. Page. 1995. Linkage Mapping of the Honey Bee (*Apis mellifera* L.) with RAPD Markers. *Genetics* 76:1371-1382.

Jansha. A. 1771. *Abhandlung von Schwärmen der Bienen.* Wien. Summarized in Anton Jansha on the swarming of bees. (Transl. by H.M. Fraser), *Apis Club,* Royston, Herts, 1951).

Kaftanoglu, O. and Y. Peng. 1980. A washing technique for collection of honey bee semen. *J. Apic. Res.* 19:205-211.

Koeniger, G. 1986. Mating sign and multiple mating in the honeybee. *Bee Wld.* 41:141-150.

Koeniger, G., N. Koeniger, M. Fabritius. 1979. Some detailed observations of matings in the honeybee. *Bee World* 34:53-57.

Koeniger, N. 1970. Factors determining the laying of drone or worker eggs by the queen honeybee. *Bee Wld.* 51:166-169.

Kühnert, M.E. and H.H. Laidlaw. 1994. Simplified apparatus for instrumental insemination of queen bees with the "Flexible insemination technique". *Apidologie* 25:144-154.

Kühnert, M, M.J. Carrick, and L.F. Allen. 1989. Use of homogenized drone semen in a bee breeding program in Western Australia. *Apidologie* 20:371-381.

Laaksonen, Milko.1987. *Mehiläishoito Emonkasvatous.* Suomen mehilaishoitajain. Lütto SML n.y. Helsinki

Laidlaw, H.H. 1934. The reproductive organs of the queen bee in relation to artificial insemination. M.Sc. thesis, *Louisiana State University,* Baton Rouge.

Laidlaw, H.H. 1939. The morphological bases for an improved technique of artificial insemination of queen bees of *Apis mellifica* Linnaeus. Ph.D. dissertation, *University of Wisconsin,* Madison.

Laidlaw, H.H. 1944. Artificial insemination of the queen bee (*Apis mellifera* L.): Morphological basis and results. *J. Morphol.* 74:429-465.

Laidlaw, H.H. 1949. Development of precision instruments for artificial insemination of queen bees. *J. Econ. Entomol.* 42:254-261.

Laidlaw, H.H. 1954. Beekeeping management for the bee breeder. *Amer. Bee J.* 94:92-95.

Laidlaw, H.H. 1956. Mates for hybrid queens. *Amer. Bee J.* 96:64

Laidlaw, H.H. 1956 (1958). Organization and operation of a bee breeding program. In: *Proceedings Tenth International Congress of Entomology* 4:1067-1078.

Laidlaw, H.H. 1974. Relationships of bees within a colony. *Apiacta* 9:49-52.

Laidlaw, H.H. 1976. *Instrumental Insemination of Queen Honey Bees.* Slide Set. Dadant and Sons, Inc. Hamilton, Illinois.

Laidlaw, H.H. 1977. *Instrumental Insemination of Honey Bee Queens.* Dadant and Sons, Hamilton, Illinois.

Laidlaw, H.H. 1979. *Contemporary Queen Rearing.* Dadant & Sons, Hamilton, Illinois.

Laidlaw, H.H. 1981. Honey bee genetics and its relationship to pollinator breeding. *Honeybee Sci.* 2:1-4. (In Japanese with an English summary.)

Laidlaw, H.H. 1987. Instrumental insemination of honeybee queens: its origin and development. *Bee Wld.* 68:17-36, 71-88.

Laidlaw, H.H. 1988. One-piece queen holder for Mackensen-type insemination device. *Amer. Bee J.* 128:281

Laidlaw, H.H. 1989. Origin and Development of Instrumental Insemination of Queen Bees. In: *The Instrumental Insemination of the Queen Bee.* R.F.A. Moritz, ed. Apimondia, Bucharest. pp 9-17.

Laidlaw, H.H. 1992. Production of queens and package bees. In: *The Hive and the Honey Bee,* J.M. Graham, ed., Dadant and Sons, Inc., Hamilton, Illinois. pp 989-1042.

Laidlaw, H.H. and J.E. Eckert. 1950. *Queen Rearing*. Dadant and Sons, Inc., Hamilton, Illinois.

Laidlaw, H.H. and J.E. Eckert. 1962. *Queen Rearing*, 2nd Edition, Revised and Enlarged. University of California Press, Berkeley, CA.

Laidlaw, H.H, F.P. Gomes, and W. E. Kerr. 1956. Estimation of the number of lethal alleles in a panmitic population of *Apis mellefera*. *Genetics* 41:179-188.

Laidlaw, H.H. and J.R. Goss. 1990. Laidlaw-Goss queen bee pre-set artificial insemination instrument. *Amer. Bee J.* 130:734-737.

Laidlaw, H.H. and C. Lorenzen. 1977. Laidlaw instrumental insemination instrument. *Amer. Bee J.* 117:428-432.

Laidlaw, H.H. and R.E. Page.1984. Polyandry in honey bees (*Apis mellifera* L.): sperm utilization and intracolony genetic relationships. *Genetics* 108:985-997.

Laidlaw, H.H. and R.E. Page. 1986. Mating designs. In: *Bee Genetics and Breeding*. T.E. Rinderer, ed., Academic Press, Orlando. pp 323-344.

Langstroth, L.L. 1853. *Langstroth on the Hive and the Honey-Bee*. The A.I.Root Company, Medina, Ohio.

Langstroth, L.L. 1888. *Langstroth on the Hive and the Honey Bee*. Revised by Dadant. Twentieth Century Edition. Dadant and Sons. Hamilton, Ill.

Larch, E.C. 1876. Grafting queen cells. *Glean. Bee Cult.* 4:48.

Léuckart, Rudolph 1861. The sexuality of bees. *Amer. Bee J.* 1:241-250.

Lush, L.L. 1945. *Animal Breeding Plans*. Iowa State College Press, Ames, Iowa.

Mackensen, O. 1943. The occurrence of parthenogenetic females in some strains of honeybees. *J. Econ. Entomol.* 36:465-467.

Mackensen, O. 1947. Effect of carbon dioxide on initial oviposition of artificially inseminated and virgin queens. *J. Econ. Entomol.* 40:344-349.

Mackensen, O. 1951. Viability and sex determination in the honey bee (*Apis mellifera* L.). Genetics. 36:500-509.

Mackensen, O. and W.C. Roberts.1948. *A Manual For The Artificial Insemination Of Queen Bees*. United States Department of Agriculture Bureau of Entomology and Plant Quarantine, ET-250,Washington, D.C.

Mackensen, O. and K.W. Tucker. 1970. *Instrumental Insemination of Queen Bees*. Agriculture Handbook No. 390, United States Department of Agriculture, Agriculutural Research Service, Washington, D.C.

Melampy, R.M. and D.B. Jones. 1939. Chemical composition and vitamin content of royal jelly. *Proc. Soc. Exp. Biol. & Med.* 41:382-388.

Miller, C.C. 1912. How best queen cells can be secured. *Amer. Bee J.* 52:243.

Moritz, R.F.A. 1983. Homogeneous mixing of honey bee semen by centrifugation. *J.Apic.Res.* 22:249-255.

Moritz, R.F.A. 1984. The effect of different diluents on the insemination success of using mixed semen. *J.Apic.Res.* 23:164-167.

Moritz, R.F.A.(ed.) 1987. *The Instrumental Insemination of the Queen Bee.* Apimondia, Bucharest.

Moritz, R.F.A. and M. Kühnert. 1984. Seasonal effects on artificial insemination of honey bee queens (*Apis mellifera* L.). *Apidologie* 15:223-231.

Nachtsheim, H. 1913. Cytologische Studien über die Geschlechts-bestimmung bei der Honigbiene (*Apis mellifica* L.). *Archiv für Zellforsch.* 11:169-241.

Nelson, J.A. 1915. *The Embryology of the Honey Bee.* Princeton University Press, Princeton, N.J.

Nelson, D.L. and H.H. Laidlaw. 1988. An evaluation of instrumentally inseminated queens shipped in packages. *Amer. Bee J.* 128:279-280.

Nolan, W.J. 1932. *Breeding the Honeybee under Controlled Conditions.* United States Department of Agriculture Technical Bulletin 326.

Oertel, E. 1930. Metamorphosis of the honeybee. *J. Morphol. Physiol* 50:295-340.

Onions, G.W. 1912. South African "fertile-worker bees". *Agr. Jour. Union S. Africa* 3:720-728.

Örösi-Pal Z. 1951. La Revue Francaise d'Apiculture. 68, special number: 281-284. (Reported by C. Toumanoff in *Les Maladies des abeilles*).

Paddock, F.B.1941. Control of American Foul Brood. *Iowa Agri. Ext. Circ.* 212 *(revised).*

Page, R.E. 1986. Sperm utilization in social insects. *Ann. Rev. Entomol.* 31:297-320.

Page, R.E. and M.K. Fondrk. 1995. The effects of colony-level selection on the social organization of honey bee (*Apis mellifera* L.) colonies: colony-level components of pollen hoarding. *Behav. Ecol. Sociobiol.* 36:135-144.

Page, R.E and N.E. Gary.1990. Genotypic variation in susceptibility of honey bees, *Apis mellifera*, to infestation by tracheal mites, *Acarapis woodi. Exp. and Appl. Acarol.* 8:275-283.

Page, R.E. and H.H. Laidlaw.1982. Closed population honey bee breeding. I. Population genetics of sex determination. *J. Apic. Res.* 21:30-37.

Page, R.E., R.B. Kimsey, and H.H. Laidlaw. 1984. Migration and dispersal of spermatozoa in spermathecae of queen honeybees (*Apis mellifera* L.). *Experientia* 40:182-184.

Page, R.E. and H.H. Laidlaw. 1992. Honey Bee Genetics and Breeding. In: *The Hive and the Honey Bee,* J.M. Graham, ed., Dadant and Sons, Inc., Hamilton, Illinois, pp 235-267.

Park, O.W. 1949. The Honey-Bee Colony-Life History. In *The Hive and the Honey Bee,* R.A. Grout, ed. Dadant and Sons, Inc., Hamilton, Illinois pp 21-78.

Parks, H.B. 1931. Bee behavior during queen succession. In *Report of the State Apiarist of Iowa for the year ending Dec.31,1931,* pp. 53-56.

Pellett, F.C. 1916. *Productive Beekeeping.* J.B. Lippincott Company, Philadelphia.

Pellett, F.C. 1918. *Practical Queen-Rearing.* American Bee Journal, Hamilton, Illinois.

Peng, Y.S., Y. Fang, S. Xu, and L. Ge. 1987a. The resistance mechanisms of the Asian honey bee, *Apis cerana* Fabr. to an ectoparasitic mite,*Varroa jacobsoni* Oudemans. *J. Invert. Pathol.* 49:54-60.

Phillips, E.F. 1928. *Beekeeping.* The Macmillan Co., New York.

Polhemus, M.S. and O.W. Park. 1951. Time factors in mating systems for honey bees. *J. Econ. Entomol.* 44:639-642.

Pritchard, M. 1932. Queen cells by the thousand. *Glean. Bee Cult.* 60:147-150.

Quinby, M. 1853. *Mysteries of Beekeeping Explained.*

Quinby, M. 1873. Bees by mail. *Amer. Bee J.* 9:37.

Réaumur, R.A.F. 1740. *Mémoires Pour Servir á l' Histoire des Insectes,* 6 vols. De l'mprimerie Royal, Paris. (cited in Huber 1814).

Ribbands, C.R. 1953. *The Behaviour and Social Life of Honeybees.* Bee Research Asssociation Limited, London.

Roberts, W C. 1944. Multiple mating of queen bees proved by progeny and flight tests. *Glean. Bee Cult.* 72:255-59,303.

Root, A.I. 1884. (Comments) *Glean. Bee Cult.* 12(19): 659-60; 12(20): 728-729.

Rothenbuhler, W.C., J.M. Kulincevic, and W.E. Kerr. 1968. Bee genetics. *Annu. Rev. Gen.* 2:413-438.

Ruttner, F. 1969. *The Instrumental Insemination of the Queen Bee.* Apimondia Publishing House, Bucharest.

Ruttner, F. 1983. *Queen Rearing: Biological Bases and Technical Instruction.* Apimondia Publishing House, Bucharest.

Ruttner, F. 1988. *Breeding Techniques and Selection for Breeding of the Honeybee.* The British Bee Breeders Association,

Ruttner, F. 1996. *Zuchttechnik und Zuchtauslese bie der Biene.* Auflage Ehrenwirth Verlag. München.

Schirach, M.A.G. 1787. *Histoire Naturelle de la Reine des Abeilles, avec l'art de former des Essaims.* Le tout Traduit de l'Allemand ou recueille par J.J. Blassiere.

Schley, P.1983. *Praktische Anleitung zur instrumentellen Besamung von Bienenköniginnen.* Selbst verlag. W. Seip,

Hauptstr. 34-36,6308 Butzbach 12, Germany.

Schley, P. 1990. *Einführung in die instrumentelle Besamung von Bienenköniginnen*. Köhler Offset K.G, Kiesweg 23, 6300 Giessen, Germany.

Shafer, G.D.1917. A study of the factors which govern mating in the honey-bee. *Michigan Agricultural College Experiment Staation Technical Bulletin 34*.

Siebold von, C.T.E. 1857. *On a True Parthenogenesis in Moths and Bees*. (Transl. by Dallas.).

Sinnott, E.W., L.C. Dunn, and T. Dobzhansky. 1950. *Principles of Genetics*. McGraw-Hill Book Company, Inc., New York.

Smith, J. 1923. *Queen Rearing Simplified*. The A.I. Root Company, Medina, Ohio.

Smith, J. 1949. *Better Queens*. Printed for J.Smith.

Smith, M.V. 1959. The production of royal jelly. *Bee Wld. 40*:250-254.

Snodgrass, R.E. 1910. *The Anatomy of the Honey Bee*. Technical Series, No. 18.Government Printing Office, Wahington, D.C.

Snodgrass, R.E. 1925. *The Anatomy and Physiology of the Honeybee*. McGraw-Hill, New York.

Snodgrass, R.E. 1956. *Anatomy of the Honey Bee*. Comstock Publishing Associates, Cornell University Press, Ithaca, New York.

Swammerdam, J. 1732. *Biblia Naturae* (A General History of Insects).

Szabo, T.I. and L.P. Lefkovitch. 1987. Fourth generation of closed-population honeybee breeding. (1) Comparison of selected and control strains. *J. Apic. Res. 26*:170-180.

Townsend, O.H. 1880. How to get plenty of choice queen cells. *Glean. Bee Cult. 8*:322-323.

Tiesler,F.-K. and E. Englert.1989. *Aufzucht, Paarung, und Verwertung von Königinnen*. Ehrenwirth Verlag GmbH München

Tucker, K.W. 1958.Automictic parthenogenesis in the honey bee. Genetics 43:299-316

Tucker, K.W. 1986. Visible mutants. In: *Bee Genetics and Breeding*, T.E. Rinderer, ed., Academic Press, Orlando, pp. 57-90.

White, E.B. 1984. *The Second Tree from the Corner*. Harper and Row, New York.

White, E.B. 1973. *Poems and Sketches of E.B.White*. Harper and Row, New York.

Watson, L.R. 1927. *Controlled Mating of Queen Bees*. American Bee Journal, Hamilton, Illinois

Woodrow, A.W. 1941. *Some effects of temperture, relative humidity, confinement, and type of food on queens in mailing cages*. United States Department of Agriculture Mimeograph Series E-529.

Woyke, J. 1955. Multiple mating of the honeybee queen (*Apis mellifica* L.) in one nuptial flight. *Bull. Acad Polon. Sci* Ch.II Vol.III(5):175-180.

Woyke, J. 1977. The hereditary color patterns in the honey bee. In: *International Symposium on Genetics, Selection, and Reproduction of the Honey Bee*, Apimondia, Bucharest, pp.49-55.

Woyke, J. 1986. Sex determination. In: *Bee Genetics and Breeding*, T.E. Rinderer, ed., Academic Press, Orlando, pp. 91-119.

Woyke, J. and Z. Jasinski. 1976. The influence of age on the results of instrumental insemination of honeybee queens. *Apidologie* 7:301-306.

Woyke, J. and Z. Jasinski. 1982. Influence of the number of attendant workers on the number of spermatazoa entering the spermatheca of instrumentally inseminated queens kept outdoors in mating nuclei. *J. Apic. Res.* 21:129-133.

Woyke, J. and Z. Jasinski. 1976 The influence of age on the results of instrumental insemination of honeybee queens. *Apidologie* 7:301-306.

Woyke, J. and Z. Jasinski. 1978. Influence of the age of drones on the results of instrumental insemination of honeybee queens. *Apidologie* 9:202-212.

Woyke, J. and F. Ruttner.1958. An anatomical study of the mating process in the honey bee. *Bee Wld.* 39:3-18.

Zander, E.1911. *Der Bau der Bien.* Verlagbuchhandlung, Eugen Ulmer, Stuttgart.

The Smoker

Three personal tools distinguish the beekeeper: veil, hive tool, and smoker. The smoker is the most important of the three in handling bee colonies. But its efficiency is directly related to the kind of smoke produced and how it is applied.

Different fuels produce different smoke. Burlap has been a standard fuel for decades. If clean, it is fairly good. Chopped punk wood, especially pine, is also good; as are pine needles, pine chips, or heavy shavings. Certainly among the very best fuels is a mixture of pine chips or partly punk wood with pine rosin bits or pine resin. This gives a dense white smoke of pleasant odor and long duration that is unexcelled in quieting the bees. In starting the smoker fire, a wad of rumpled newspaper is lighted and dropped into the fire chamber. While working the bellows, the chamber is partly filled with charcoal of previous smoker use or with chips or small chunks of punk wood. When the wood is burning well, add more wood and about one tablespoon of the pine rosin bits on top of the wood. Chips can be added as needed later, followed by small amounts of rosin.

Charcoal for igniting the smoker for its next use is automatically acquired if the smoker is extinguished by a wad of newspaper in the nozzle. Before the smoker cools, lift the firebox lid by prying with the hive tool until the back edge clears the fire chamber; otherwise it may be permanently stuck!

One caveat: the rosin smoke coats the firebox walls with tar. Fortunately, by building a fire in the firebox and getting a good flame the tar burns to crisp and is easily knocked off the walls by tapping the firebox with the hive tool., leaving the firebox clean, When the tar is burning well, lower the smoker lid part way over the firebox so the tar in the lid will be burned.

The Superhorse

"Beekeepers' back"—a phrase commonly heard among beekeepers—seems to have some connection with lifting the heavy bodies of bee hives. A device (Fig. 115), like a double sawhorse, used by some beekeepers in south Florida for decades, eliminates much of the strain of handling supers that are on the hive. With the superhorse placed to the back and side of the beekeeper, the upper hive bodies—and usually the heaviest—can be set on the

Figure 115. The Superhorse.

Figure 116. Smoker and nucleus seat.

superhorse by turning partly around. The usually lighter bodies from near ground level can be examined and rearranged on the superhorse at a comfortable height rather than by stooping.

Incidentally, a smoker between the legs is a fetter.

The Nucleus Seat

Nuclei are usually set on the ground in queen mating apiaries. In working the nuclei, the queen rearer must stoop or kneel on one knee or on both knees. The nucleus seat (Fig. 116) minimizes the strain on the beekeeper's back and also is also a carryall for queen cages, caged queens, and queen rearing paraphernalia.

The Robber Screen

Robbing is an ever-present possibility in apiaries of more than one colony. Incitements to robbing are bits of comb with honey or other scrapings from the hive left on the ground by the hive, a hive opened too long when foragers find little or no food in the field, or when the colonies in the apiary have been, or are, undergoing stress such as rough handling or vandalism. The need to prevent robbing is obvious in the these situations.

If robbing is more vigorous than merely some bees trying to enter the entrance of a colony and are being intercepted by the colony guard bees, or bees are investigating possible entrance through cracks or holes in the hive body, the situation can become disastrous to the whole apiary. Reducing the hive entrance to a small opening with blocks or a v-screen may cause the colony to overheat as the robbers block ventilation and bee passage. Robbing can be positively controlled by equipping each colony in the apiary with a "robber screen" that fits on the hive front over the entrance (Fig. 117). Robbers cannot seriously reduce the colony ventilation, and the colony bees defend the colony with the advantage of the protection of the screen.

A robber screen is a frame that is covered on the outer side with course screen. It permits the colony bees to come out of the hive entrance and move about on the hive front. The frame has a three-eighths inch space along the top between the frame and the hive which can serve as the main hive entrance. The robber screen covers all but about one or two inches of the bottom-board open entrance. A small opening at one end of the screen is left for the hive entry of a supersedure queen returning from a flight. She may not find the entrance at the top of the screen. The workers like the lower entrance, too, but as it becomes crowded the flight from the top increases.

The colony bees will use the unscreened entrance and the top opening. The robbers will try to enter the hive at the points where the hive odor is strong, that is, through the screen, and through the small open entrance. With the protection of a robber screen the colony bees can defend the colony under even the most severe robber attacks, and robbing will not occur. Within 15 to 30 minutes efforts to rob a colony will have ceased, and only a few probing, but unsuccessful, robbers will be in evidence. Robber screens placed on the entrances of colonies on pallets can direct the flight of bees to the outer sides of the entrances to lessen drifting from one hive to the other.

Figure 117a. Outer side of robber screen

Figure 117b. Inner side of robber screen

The Role of the Drone

The essentiality of drones to the life of a honey bee colony is obscured by the apparent uselessness of drones that are visible in a colony, and the invisibility of the true fathers of the colony's workers. The true fathers are represented by their sperm in the spermatheca of the queen mother. Through their sperm, the drones' reproductive-lives are extended to that of their queen mate. This preservation of sperm beyond the body-lives of the males producing them predates by an eon the cryogenic preservation of mammalian sperm.

Sperm preservation in the spermatheca is necessary for colony formation and existence. The queen's mating to fertilize each egg as it is laid would handicap oviposition to such an extent that colonies of more than a few workers would be impossible.

A drone is genetically a gamete of his mother, and his own gametes are clones of his genetic composition. This makes a drone's mother the genetic "father" of her son's progeny, and allows mother-daughter matings. A population of drones in a colony reflects the genetic homozygosity-heterozygosity of the queen mother of the colony. It does not indicate the queen's mating.

The active role of drones in brood rearing brings up the question of whether drones form a caste of honey bees. The definition of "caste" in social insects, as given in several biological dictionaries, is essentially that—*any set of individuals that are both mor-*

Figure 118. Product of queen rearing and bee breeding—a vigorous well-bred laying queen.

phologically distinct and specialized in behavior, in a given colony, is a caste. The queen, workers and drones fit this definition. They are each derived from a matured egg of the queen mother with the destiny of the mature egg environmentally determined.

Incidentally

Smoking many of the older bees up into a cluster box above the brood nest, and setting the cluster box off on an inner cover before examining the brood nest combs, lessens stings when examining colonies of difficult bees.

SONG OF THE QUEEN BEE

"The breeding of the bee," says a United States Department of Agriculture bulletin on artificial insemination, "has always been handicapped by the fact that the queen mates in the air with whatever drone she encounters."

E. B. White

When the air is wine and the wind is free
And the morning sits on the lovely lea
And sunlight ripples on every tree,
Then love-in-air is the thing for me—
* I'm a bee,*
* I'm a ravishing, rollicking, young queen bee,*
* That's me.*

I wish to state that I think it's great,
Oh, it's simply rare in the upper air,
* It's the place to pair*
* With a bee,*
Let old geneticists plot and plan,
They're stuffy people, to a man;
Let gossips whisper behind their fan.
* (Oh, she does?*
* Buzz, buzz, buzz!)*
My nuptial flight is sheer delight;
I'm a giddy girl who likes to swirl,
* To fly and soar*
* And fly some more,*
*   I'm a bee.*
And I wish to state that I'll always mate
* With whatever drone I encounter.*

There's a kind of a wild and glad elation
In the natural way of insemination;
Who thinks that love is a handicap
Is a fuddydud and a common sap,
For I am a queen and I am a bee,
I'm devil-may-care, and I'm fancy-free,

The test tube doesn't appeal to me,
  Not me,
  I'm a bee,
And I'm here to state that I'll always mate
  With whatever drone I encounter.

Let mares and cows, by calculating,
Improve themselves with loveless mating,
Let groundlings breed in the modern fashion,
I'll stick to the air and the grand old passion;
I may be small and I'm just a bee
But I won't have Science improving me,
  Not me,
  I'm a bee.
On a day that's fair with a wind that's free,
Any old drone is the lad for me.

I have no flair for love moderne,
It's far too studied, far too stern,
I'm just a bee—I'm wild, I'm free,
  That's me.
I can't afford to be too choosy;
In every queen there's a touch of floozy,
  And it's simply rare
  In the upper air
  And I wish to state
  That I'll always mate
With whatever drone I encounter.

Man is a fool for the latest movement,
He broods and broods on race improvement;
What boots it to improve a bee
If it means the end of ecstasy?
     (He ought to be there
  On a day that's fair,
  Oh, it's simply rare
   For a bee.)
Man's so wise he is growing foolish,
Some of his schemes are downright ghoulish;
And he owns a bomb that'll end creation
And he wants to change the sex relation,
He thinks that love is a handicap,
He's a fuddydud, he's a simple sap;
Man is a meddler, man's a boob,
He looks for love in the depths of a tube,
His restless mind is forever ranging,
He thinks he's advancing as long as he's changing,

He cracks the atom, he racks his skull,
Man is meddlesome, man is dull,
Man is busy instead of idle,
Man is alarmingly suicidal.
  Me, I'm a bee.

I am a bee and I simply love it,
I am a bee and I'm darned glad of it,
I am a bee, I know about love:
You go upstairs, you go above,
You do not pause to dine or sup,
The sky won't wait—it's a long trip up;
You rise, you soar, you take the blue,
It's you and me, kid, me and you,
It's everything, it's the nearest drone,
It's never a thing that you find alone.
  I'm a bee,
  I'm free.

If any old farmer can keep and hive me,
Then any old drone may catch and wive me;
I'm sorry for creatures who cannot pair
On a gorgeous day in the upper air,
I'm sorry for cows who have to boast
Of affairs they've had by parcel post,
I'm sorry for man with his plots and guile,
His test-tube manner, his test-tube smile;
I'll multiply and I'll increase
As I always have—by mere caprice;
For I am a queen and I am a bee,
I'm devil-may-care and I'm fancy-free,
Love-in-air is the thing for me,
  Oh, it's simply rare
  In the beautiful air,
    And I wish to state
    That I'll always mate
With whatever drone I encounter.

Admonition

Lawrie Overton Willis
Room 402-717 Market St.
San Francisco 3, Calif.

From   SAN FRANCISCO CHRONICLE, December 5, 1948:

Davis, Dec.4—Development of a bee of good nature and great honey-producing qualities is under way by artificial insemination in the College of Agriculture on the Davis campus of the University of California. The new inseminator, designed by Dr. Harry H. Laidlaw, assistant professor of entomology, is declared superior to any similar instrument heretofore developed. By its use a queen bee can be put to sleep with carbon dioxide gas and inseminated from a selected drone in five minutes.

Poor, poor Queen!   How could you!!

L.O.W.

Response

The queen bee is a giddy thing
She plays, she romps, she takes the wing
She darts among the sun-lit hives
Without a thought of being wise.

Her offspring slave throughout the day
They feed her children as best they may.
They would like to see a movement
Directed toward stock improvement.

H.H. Laidlaw, Jr.

Index

R

Y

Yeasts 48
Yugo bees 171, 195-196
Yugoslavian bees 171, 195

Z

Zander, E. 127
Zygote 26, 142, 150